AF291555

# THE
# SOUTHERN
# RAILWAY
# 1923–1935

# THE SOUTHERN RAILWAY 1923–1935

## THE FORMATIVE YEARS

CHARLES PHILLIPS

PEN & SWORD TRANSPORT

AN IMPRINT OF PEN & SWORD BOOKS LTD.
YORKSHIRE – PHILADELPHIA

First published in Great Britain in 2026 by
Pen and Sword Transport
An imprint of
Pen & Sword Books Ltd.
Yorkshire – Philadelphia



ISBN 978 1 03610 472 6



Typeset in 12/16 Palatino by SJmagic DESIGN SERVICES, India.

Printed and bound in India by Replika Press Pvt. Ltd.

The Publisher's authorised representative in the EU for product safety is Authorised Rep Compliance Ltd., Ground Floor, 71 Lower Baggot Street, Dublin D02 P593, Ireland.
www.arccompliance.com

For a complete list of Pen & Sword titles please contact

PEN & SWORD BOOKS LIMITED
George House, Beevor Street, Off Pontefract Road, Hoyle Mill, Barnsley, South Yorkshire, England, S71 1HN.
E-mail: enquiries@pen-and-sword.co.uk
Website: www.pen-and-sword.co.uk

or

PEN AND SWORD BOOKS
1950 Lawrence Rd, Havertown, PA 19083, USA
E-mail: uspen-and-sword@casematepublishers.com
website: www.penandswordbooks.com

# CONTENTS

# INTRODUCTION

The Southern Railway was the smallest of the 'Big Four' railway companies created under the Railways Act 1921 following the First World War. With a total route mileage of 2,129 miles it was smaller than the Southern Division of the London and North Eastern Railway whose total route mileage was 3,095 miles. Its area stretched from Kent in the east to Cornwall in the west and Devon in the south and London in the north. With the Isle of Wight railways, it was the only one of the 'Big Four' to operate a railway system detached from the railways of mainland Britain. At its formation, the Southern Railway was a railway of contradictions in that on the Isle of Wight it had steam locomotives dating back to the mid-1860s, but in the London area there were electric trains operating off the direct current third rail system and the alternating current overhead system. Unlike the other three groups, the Southern Railway's main revenue came from passengers as opposed to freight and because of this the Company heavily invested in electrification on the direct current third rail system, replacing those sections using the alternating current overhead system. As to whether adopting the direct current third rail system was wise is still a matter for debate, but by doing so it did enable the electrification to be done more quickly and more cheaply than would otherwise have been the case.

This book covers the years 1923, which saw the formation of the Southern Railway, to 1935, which saw the completion of the first stage of the Company's main line electrification, the building of the new Western Docks at Southampton and the Company beginning its involvement in air travel in the form of Railway Air Services. Some new lines opened, such as the Wimbledon and Sutton, but other lines either lost their passenger service (such as the Canterbury and Whitstable) or closed totally, such as the Lynton and Barnstaple. It was a period of unrest and also tragedy with the General Strike in May 1926 and the Sevenoaks accident in August 1927. The Company's motive power policy during these years saw the building of new versions of some existing classes from the London and South Western and South Eastern and Chatham railways, but also totally new classes in the form of the Lord Nelson Class and the Schools Class and whilst the former had the misfortune not to be built in sufficient numbers due to the Company having the 'electrification bug', the latter class proved to be an outstanding success.

I like the Southern Railway and decided to write its history because I used to work with a lot of people who lived in places served by it and travelled by it, but it was similar to the former Great Eastern Railway in terms of its main revenue. In writing this history, I have used previously printed histories, but also contemporary newspapers and magazines such as the *Railway Magazine*, the *Locomotive Magazine* and *Railways*. I wish to thank the Great Eastern Railway Society making the latter two magazines available on CD.

The only former Southern Railway official that I ever met was A B Macleod when I was researching photographs in the Ian Allan archive for my book 'The Great Eastern Since 1900'.

Charles Phillips

# Chapter 1

# IN THE BEGINNING 1923 TO 1926
## THE FIRST DAY

The Southern Railway (SR) was the smallest of the four big railway companies formed by the Railways Act of 1921. It comprised the London and South Western Railway (LSWR), the London, Brighton and South Coast Railway (LBSCR), the South Eastern Railway (SER) and the London, Chatham and Dover Railway (LCDR), the latter two of which had been working together in a working union since 1899 as the South Eastern and Chatham Railway (SECR). There were additionally some smaller companies absorbed into the new company. These were: the Isle of Wight Central Railway (IWCR), the Isle of Wight Railway (IWR), the Freshwater, Yarmouth and Newport Railway (FYNR), the Lynton and Barnstaple Railway (LBR), the Plymouth, Devonport and South Western Junction Railway (PDSWJR) and other smaller concerns which did not own their own rolling stock. Additionally, the company owned the Somerset and Dorset Railway (SDR) jointly with the London, Midland and Scottish Railway (LMS). It participated jointly with the Great Western Railway (GWR) in the Weymouth and Portland and the Easton and Church Hope railways, with the GWR and the LMS in the West London Extension Railway and with the London and North Eastern Railway (LNER) and the Metropolitan (MR) and Metropolitan District (MDR) railways in the East London Joint Committee.

According to *the Daily News* of Monday 1 January 1923:

The last night of the old year was one of hectic activity for many thousands on the railway staffs. Columns on columns of accounts were being balanced, and stacks of tickets prepared for issue. The four groups whose creation should ultimately conduce to higher efficiency and economy in management are: The Southern Railway (2,129 miles), including the London and South Western, South Eastern and Chatham, and London and Brighton and South Coast lines.

The sixteen general managers will be reduced to four. So far only three of them have been appointed. They are: Mr. Arthur Watson. London. Midland and Scottish. Mr. Felix J. C. Pole. Great Western. Mr. R. L. Wedgewood, London and North Eastern. No manager of the Southern Group has yet been appointed. But in unofficial railway circles the name of Sir Herbert Walker. General Manager of the London and South Western Railway is freely mentioned as a most likely' selection. It is not yet decided whether Waterloo or London Bridge will become the headquarters of the new Southern Railway, but the others will be: London and North Eastern Railway. Marylebone: London Midland and Scottish Railway. Euston; Great Western Railway. Paddington.

## A NEW LARGER RAILWAY

The new company operated passenger services from Ramsgate in the east to Padstow in the west and from London in the north to Plymouth in the south. It was a participant in a through service from the northwest of England through to the south coast and Kent. Titled as the Sunny South Express by the new company, SR motive power took over at Kensington, Addison Road (now Olympia) for the last part of the journey to the coast and hopefully plentiful sunshine. Through the Somerset and Dorset Railway it also participated in another through service from the north east and north west of England to Bournemouth. Originating in Midland Railway and LSWR days, in 1927 it was elevated to named status as the Pines Express.

The company exchanged goods traffic with the other three big companies in the London area except for the GWR where the exchange was made at Reading.

Whilst the number of staff inherited by the SR is not known, according to the *Railway Year Book* for 1924, on 21 March 1923 the company employed 70,479 staff.

Of the three principal companies grouped to form a new larger SR, the LSWR was both the largest and arguably the best. Its main lines ran from London Waterloo Station to Salisbury, Exeter and Plymouth, Portsmouth, Southampton, Bournemouth, and by running powers over the GWR to Weymouth. There was an outer suburban line to Reading, the important Berkshire town being reached by running powers over the SECR. Other principal outer suburban destinations from Waterloo included Guildford, Windsor and Epsom. Longer distance passenger traffic included ocean liner boat trains and also cross Channel and Channel Islands boat trains from the major port of Southampton.

As well as a large London suburban service, much of which had been electrified on the third rail dc system, it also carried holiday makers to the Isle of Wight, Bournemouth and other resorts in Hampshire, Dorset, Devon and Cornwall. There was also lucrative race traffic to Ascot, Sandown Park and Kempton Park. It also served the Naval depots at Portsmouth, Portland and Devonport, the Army depots at Aldershot and on Salisbury Plain and the Air Force establishment at Farnborough. Its freight included Cornish china clay, imports of all sorts through Southampton, domestic agricultural produce from all over its system as well as coal and manufactured articles from the North and the Midlands via its marshalling yard at Feltham. The company's packet steamers served the busy ports of Le Havre, St Malo, Cherbourg, Guernsey and Jersey and jointly with the LBSCR from Portsmouth to Ryde on the Isle of Wight. Often playing second fiddle in railway history to the GWR, the LSWR was in reality a giant in its own right, one which had embraced electrification and even operated a bus service from Exeter to Chagford and jointly with the GWR from Weymouth to Wyke Regis.

The company's stations were a bit of a mixed bag. The flagship of the LSWR and later, a new SR, was always Waterloo, the complete rebuilding of which had been only recently finished in 1922 after over a decade's work, much of which had been interrupted by the Great War. Bournemouth Central was equally good, but others such as Exeter (Queen Street), Woking and Southampton West left a great deal to be desired and needed urgent modernisation. The LSWR had been ahead of the game in its use of lock-and-block signalling

**Waterloo Station** the headquarters of the London and South Western Railway became the headquarters of the Southern Railway. The station had been rebuilt in the early part of the twentieth century. Here is a view of Platforms 1 to 4 c.1910 at what appears to be summer holiday time. *Commercial post card*

**By comparison** this view taken of the same platforms in about 1922 show them to be almost deserted and would suggest that it was taken some time during the week between the peak hours. The new electric trains predominate. *Commercial post card*

and in 1901 had installed semi-automatic low-pressure pneumatic worked signals on the four track section of line between Woking and Basingstoke.

There was much to commend the company's carriage stock. From the turn of the century the LSWR had embarked on an ambitious construction programme of bespoke bogie stock. For the new electric services which had commenced in 1915, the bodies of old four- and six-wheeled carriages were placed on new bogie frames, the trains comprising three-car multiple units. The LSWR's motive power was a bit of a mixture. On one hand there were 2-4-0 well tanks, some 0-6-0 saddle tanks and some 0-6-0 tender locomotives dating back to the mid-1870s and the Beatties, whilst on the other hand there were also modern 4-6-0 tender locomotives and some 4-6-2 and 4-8-0 side

**This view** of Waterloo Station in the early 1920s is taken from the signal box and shows a variety of locomotives from the designs of Adams, Drummond and Urie. From the evidence of the Urie locomotive's headcode, the train would appear to be an ocean special to Southampton Docks. *Photomatic*

tanks dating from 1914 to 1922 under its final CME, Robert Urie. In addition there were various 4-4-0s, 0-6-0s, 0-4-2s, 4-6-0s, 4-2-2-0s, 0-4-4 side tanks, 4-4-2 side tanks,0-6-0 side tanks, 0-4-0 side and saddle tanks and even a 2-2-0 side tank design used for 'rail-motor' duties. Not forgetting the fearsome Dugald Drummond's own personal 4-2-4 inspection saloon, 'The Bug'.

The Company's works were at Eastleigh to where the Company had moved them in 1910,

The LSWR used the vacuum brake.

Its signal boxes were a mixture of Saxby and Farmer designs and its own designs, whilst its lever frames were a mixture of Saxby and Farmer and Stevens and Sons designs.

*Above*: **Drummond's 4-cylinder** 4-6-0s whilst looking impressive were not great performers. His T14 Class dating from 1911 were the best of his 4-6-0s and here is one following modifications by Urie and Maunsell on a Plymouth train. *H.C. Casserley*

*Below*: **In contrast** to Drummond's 4-cylinder 4-6-0s, Urie's 2-cylinder 4-6-0s were a success and three classes were built. The mixed traffic H15 coming out in 1914, the express N15 in 1918 and the goods S15 in 1920. Here is No 750 of the N15 class. *Photomatic*

***Above***: **The LSWR** could have used eight coupled goods locomotive, but instead chose to use six coupled locomotives of which the S15 4-6-0s was the most powerful class. Here is No 507 in Southern Railway days. The locomotives were mostly allocated to the London area. *Photomatic*

***Below***: **The LSWR** did not possess many named locomotives, the only ones being the twenty-five members of classes B4, D6 and K14 0-4-0Ts built between 1891 and 1908 and used for a variety of shunting jobs. Here is No 747 DINARD of Class K14 built in 1908. *John Scott-Morgan Collection*

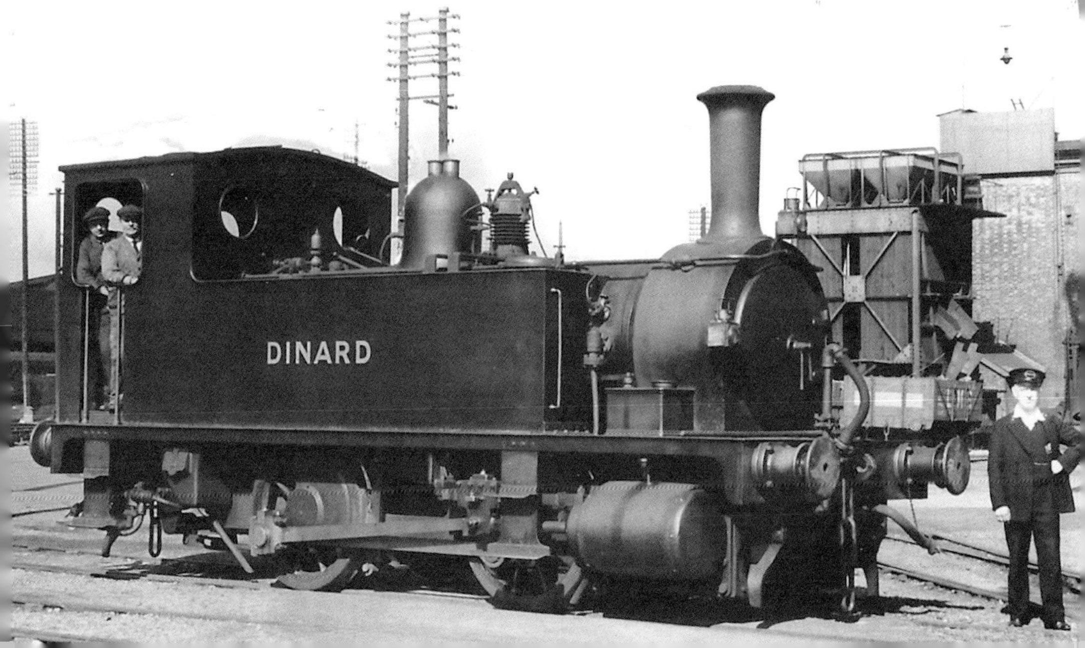

**This a** view of Basingstoke locomotive shed in about 1925. The shed was a three road shed and in the far recesses of the shed can be seen a tank locomotive. Basingstoke had an allocation of twenty-five locomotives. *H.C. Casserley*

By the end of 1922, the company had already created an extensive 600 volts dc electric suburban system of lines all operated by electric multiple units (emus). Radiating from Waterloo, this included in the first part of the ambitious programme: the Kingston Roundabout and East Putney to Wimbledon; the Hounslow Loop; Malden and Hampton Court; Strawberry Hill; and Teddington to Shepperton. There was also a short extension from Hampton Court Junction to Claygate on the section of line from Hampton Court Junction to Guildford via Cobham, which saw operation from 20 November 1916 to 31 May 1919. In the planned second stage, the electrification was to be extended from Hampton Court Junction to Guildford via Woking and Hampton Court Junction to Guildford via Cobham, as well as

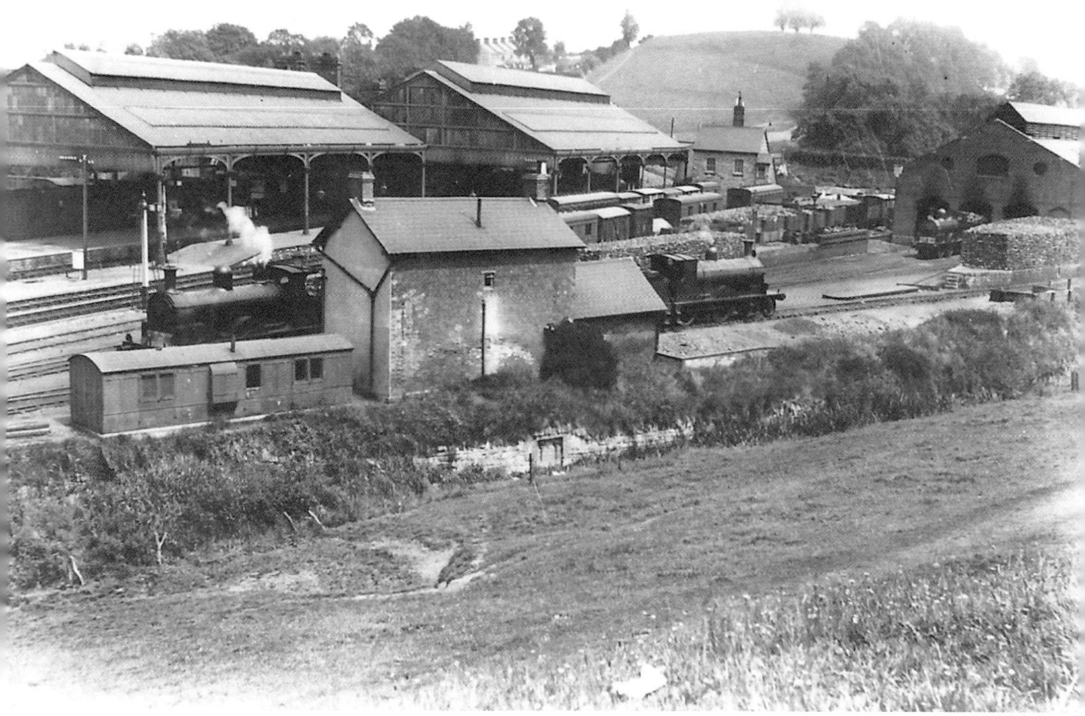

*Above*: **This photograph** taken in about 1925 shows both Yeovil Town Station, but also Yeovil locomotive shed. On the shed can be seen two A12/O4 0-4-2s and one C8 4-4-0. Class C8 were Drummond's first 4-4-0s for the LSWR whilst the A12/O4 were the last 0-4-2s to see service in Britain. *H.C. Casserley*

*Below*: **Durnsford Road** Power Station, Wimbledon in about 1920 showing an Adams T1 0-4-4T No 77. Class T1 first appeared in 1888 and were used on suburban passenger work. No 77 remained in service until withdrawn in 1932. *LSWR Official*

Raynes Park to Effingham Junction via Epsom. Other overlooked sections of the former company's inner London routes included the section from Studland Road Junction just east of Ravenscourt Park to Gunnersbury, from South Acton to Gunnersbury and from Gunnersbury to Richmond. The LSWR did not provide the main passenger service, with trains to Richmond from Hammersmith in the hands of MDR electric trains and from Willesden Junction and Broad Street, London and North Western Railway (LNWR) electric trains. The LSWR had ceased running its trains from Addison Road and on to Waterloo via Latchmere Junction after 5 June 1916 and had never run trains to South Acton over the joint LSWR and LNWR, North and South Western Junction Railway. Because the MDR used a third and fourth rail electrical system, which was also used by the LNWR, the LSWR adopted a similar system but only using three rails with current return achieved through the running rails rather than a fourth rail as was standard MDR practise. An impressive web of

**The London** and South Western Railway commenced the electrification of its London suburban lines in 1915 using three-car emus which incorporated the bodies of former steam stock carriages. Here is a six-car train formed of two units c.1916. *Commercial post card*

*Above*: **The sections** of the LSWR from Putney Bridge to Wimbledon and Ravenscourt Park to Richmond were also served by Metropolitan District Railway's electric trains and here is a train at Southfields on the Wimbledon section c.1910. *Commercial post card*

*Below*: **Basingstoke Station** on the LSWR's main line out of Waterloo was the junction for the main lines to Southampton, Bournemouth and Weymouth and Salisbury, Exeter and the West Country. This photograph of the station was taken c.1920. *Commercial post card*

*Above*: **Clandon Station** on the New Guildford Line from Effingham Junction to Guildford in about 1924 with Class M7 0-4-4T No E674. Some of the wagons in the station retain their pre-grouping initials; both G C and G E can be made out. *H.C. Casserley*

*Below*: **In comparison** to Basingstoke and Clandon stations Watergate Halt on the North Devon and Cornwall Junction Light Railway section was rather lacking in amenities. Although the Railway was authorised in 1914, it did not open until 27 July 1925. *John Scott-Morgan collection*

*Above*: **The London** and South Western Railway was an early user of low-pressure pneumatic power signalling and amongst other places equipped the section of main line between Basingstoke and Woking, which is seen in this photograph. *John Scott-Morgan collection*

*Below*: **The North** Cornwall Line from Halwill in Devon to Padstow in Cornwall was completed in March 1899 and was the most westerly line of the LSWR. This photograph is of a neat three-span girder bridge over the River Camel near Padstow. *Commercial postcard*

sparking lines, but this however was only the beginning, with plans to extend the 'juice' rails to Ascot, Windsor, Weybridge, Farnham and Aldershot and an alternative route to Guildford.

Not to be forgotten, there was also the extremely busy short underground line from Waterloo to Bank station in the City of London. This was the oldest electrified section of the SR, having been opened on 8 August 1898 and electrically worked from the start.

The SECR was formed of two formerly deeply antagonistic companies, the SER and the LCDR, with them finally burying the hatchet in the form of a new working union formed in 1899. Its main lines ran from multiple termini in London at Cannon Street, Charing Cross, London Bridge, Holborn Viaduct and Victoria to Canterbury, Margate, Ramsgate, Dover, Folkestone and Hastings as well to Reading. Other principal places served were Gravesend, Chatham, Rochester, Ashford, Tonbridge and Tunbridge Wells. Its passenger traffic included cross Channel boat trains to both Dover and Folkestone. The railway boasted a large London suburban service entirely operated by steam despite plans to electrify it at 1500 volts dc using the third rail current collection system. Away from London, the railway carried holiday makers to the seaside resorts of Margate, Ramsgate, Broadstairs, Folkestone and Hastings.

The SECR also served the Army depot at Aldershot, the Naval depot at Chatham and the Air Force establishment at Manston. Its freight included Continental traffic through Folkestone and Dover, coal including that from the Kent coalfield, industrial materials received from the northern railways, agricultural produce and livestock. The railway's ships served Boulogne and, Calais. Ships of the Belgian Marine sailed from Dover to Ostend, with the Zeeland Steamship Company operating to Flushing from Folkestone Harbour. In addition, ships of the Batavier Line sailed from Gravesend to Rotterdam.

***Above*****: The SS** *Biarritz* was constructed in 1914/5 for the SECR by Denny's of Dumbarton and on completion was taken over by the Admiralty and used for high speed mine laying prior to being returned to the SECR in 1921 for use on the cross Channel service. *Commercial postcard*

***Below*****: The TSS** *Victoria* was built for the South Eastern and Chatham Railway by William Denny and entered service in April 1907. It is seen at Boulogne c.1923. The ship was sold to the Isle of Man Steam Packet Company in 1928. *John Scott-Morgan collection*

The SECR was unique among the three constituent companies in owning a canal – the Gravesend and Rochester Canal, originally the Thames and Medway Canal. Opening between 1801 and 1824, it had originally run from the Thames at Gravesend to the Medway at Strood. In 1845 it was purchased by the Gravesend and Rochester Canal and Railway Company, who took the portion from Higham Station to Rochester for their railway track bed and renamed the remaining portion the Gravesend and Rochester Canal. In 1846, the Railway Company was taken over by the SER. The surviving length of the Canal from the tidal basin at Gravesend to Higham Station was four miles.

The railway's stations greatly varied in quality. The SECR's left hand side of Victoria station in London had been rebuilt, being finally completed in 1908. Following the collapse of the roof of Charing Cross station in 1905, a new station would emerge from

**New Romney** and Littlestone was the terminus of a branch line from Appledore and Lydd. The other arm of the branch line from Lydd served the lonely terminus at Dungeness. H Class 0-4-4T No 162 is seen at New Romney station in about 1910. *John Scott-Morgan collection*

the ruins. Dover Marine station had also been transformed by 1915 and had opened that year for military traffic, but because of the war would not cater as always intended for peacetime day trippers and continental tourists until 1919.

The quality of passenger rolling stock also greatly varied. Whilst there were quality bogie carriages for crack continental services connecting with Channel Packet steamers, including luxury Pullman cars, there were also decrepit six- and four-wheeled carriages to endure when travelling through the hop fields of Kent on a SECR branch service. The company could always be trusted to rustle up the worst examples for hop pickers' specials for Cockneys from South London heading to the Kentish Weald in the early autumn. Of the bogie carriages, few had gangway connections to other carriages. Still adhering to continental practice, the company's trains conveyed 1st, 2nd and 3rd class carriages unlike the other two companies which had already abolished 2nd class carriages. One of the first moves of a new SR was to bring the SECR into line on class provision, with all 2nd class accommodation on domestic services abolished after 1 October 1923.

The railway's motive power comprised various designs of 0-6-0, 4-4-0 and 2-6-0 tender locomotives, 0-4-0 crane tanks and saddle tanks, 0-6-0 side and saddle tanks, 0-4-4 and 0-6-4 side tanks, a 2-6-4 side tank and some steam 'rail motors' which comprised 0-4-0 side tanks plus a four-wheeled bogie carriages attached to the locomotive.

The Company's works were at Ashford, which had been opened in 1847 by the South Eastern Railway and to which the machinery and equipment of the London, Chatham and Dover Railway's works at Longhedge had been transferred in 1911.

Like the LSWR the SECR used the vacuum brake.

The Company's signal boxes were of its own design as were its lever frames.

The LBSCR's main lines ran from termini in London at London Bridge and Victoria to Brighton, Eastbourne, Epsom, Hastings,

*Above:* **The L Class** of the South Eastern and Chatham came out in 1914 with twelve members being built by Beyer Peacock of Manchester and 10 members by Borsig of Berlin. Here is No 770 built by Beyer Peacock on a train of six-wheeled stock. *John Scott-Morgan collection*

*Below:* **As the** L Class was restricted as to what routes they could work, in 1919 and 1921 respectively, Maunsell rebuilt some members of classes E and D to E1 and D1. Here is E1 No 506 on an express in about 1923. *John Scott-Morgan collection*

*Above*: **Wainwright's C** Class 0-6-0s dating from 1900 were some of the most useful locomotives that were built by the SECR, and one lasted in service until 1967. No A223 is seen on a train of six-wheelers in early Southern days. *John Scott-Morgan collection*

*Below*: **Class P** 0-6-0Ts were amongst the smallest of the South Eastern and Chatham Railway's locomotives. In about 1923 No 323 is seen at Sevenoaks on a train to Otford composed of former LCDR six wheeled carriages. No 323 is preserved on the Bluebell Railway. *Photomatic*

Horsham Portsmouth and Tunbridge Wells. Its passenger services included boat trains to Newhaven, race goers' specials to Goodwood and Epsom, as well serving the popular holiday destinations of Brighton, Hove, Worthing, Eastbourne, Hastings, Littlehampton and Portsmouth for the Isle of Wight. There was a large London suburban traffic, some of which lines had been electrified using advanced 6,600 volts ac overhead power collection and the LBSCR had plans to extend the electrification to Brighton and Eastbourne. It served the home of the Royal Navy at Portsmouth. Its freight included Continental traffic through Newhaven, coal and industrial materials received from the northern railways and agricultural produce and livestock. The railway operated ships jointly with the French State Railway (ETAT) between Newhaven and Dieppe and jointly with the LSWR between Portsmouth to Ryde.

**The London,** Brighton and South Coast Railway used large tank locomotives for many of its principal services, of which the L Class 4-6-4Ts dating from 1914 were the largest locomotives. Here is the class leader No B327 built in 1914. *John Scott-Morgan collection*

**The London,** Brighton and South Coast Railway's largest passenger tender locomotives were the eleven members of Classes H1 and H2 4-4-2s built between 1905 and 1912. Here is No B37 later 2037 *Selsey Bill* built in 1905 and named in March 1926. *John Scott-Morgan collection*

The company was progressive when it came to signalling affairs, using Sykes lock and block system and equipping its distant signals with a distinctive illuminated white fishtail beside the red lamp – the Coligny-Welch lamp for added reassurance to drivers.

The quality of the company's stations varied. Victoria, Brighton and Eastbourne were excellent, having all been recently remodelled and improved before the First World War. The less said about the company's headquarters station at London Bridge the better.

Its carriages were also something of a hotchpotch, ranging from ancient four- and six-wheelers to sumptuous new bogie eight-wheelers. The company had a penchant for rather fine Pullman cars

**The London,** Brighton and South Coast Railway stuck to tender locomotives for its main line goods services and here is No B343 one of its K Class 2-6-0s on such a working. A 2-6-2T version of the class was considered but was rejected. *John Scott-Morgan collection*

**The most** famous London, Brighton and South Coast Railway locomotives were the B1 or Gladstone Class 0-4-2s dating from 1882. Locomotive No B183 formerly Eastbourne was built in November 1889 and withdrawn in January 1929. *H.C. Casserley*

and quite a few trains, particularly on its flagship line from London to Brighton, included them in a rake or were complete Pullman sets, a fine sight with an umber Brighton Atlantic at the front heading to Brighton or Eastbourne. The Pullman cars came in various classes, composite 1st and 3rd or all 1st only or 3rd only. With its main locomotive works at Brighton, motive power comprised 0-4-2, 4-4-0, 4-4-2, 0-6-0 and 2-6-0 tender locomotives, 0-4-2, 0-4-4, 4-4-2. 0-6-0, 0-6-2 and large 4-6-2 side tanks and 4-6-4 side and well tanks. The company tended to prefer tank locomotives, given the generally shorter distances run when compared to other railways. There had even been an abortive proposal for a 2-6-2 side tank for goods traffic, whilst one of the 4-6-4 side and well tanks had originally been intended to be a 4-6-0 tender locomotive, but was instead completed as a 4-6-4 tank locomotive.

The Company's locomotive works were at Brighton and had been established in 1840 and originally also dealt with carriages and

**This is** an interior photograph of the Brighton Locomotive Works of the London, Brighton and South Coast Railway taken in about 1922. The locomotive nearest the camera is a C2X 0-6-0 The class was a rebuild of the C2 Class. *John Scott-Morgan collection*

**Portsmouth Harbour** Station opened in 1876 and was jointly owned by the London and South Western and London, Brighton and South Coast Railways. This view taken from the seaward side shows the station and pier with ships of the Portsmouth to Ryde service. *Commercial postcard*

wagons prior to the opening of Lancing carriage and wagon works in 1912.

Unlike the other two principal companies the LBSCR used the Westinghouse air brake.

Its signal boxes were a mixture of Saxby and Farmer design and its own designs, whilst its lever frames were also a mixture of Saxby and Farmer and its own design.

Of the other companies, the IWCR served Ryde, Newport, Cowes and Ventnor, the IWR served Ryde, Shanklin, Ventnor and Bembridge and the FYNR served the places in its name. Their principle passenger traffic was holiday makers and local travellers. Goods traffic was coal, industrial materials and minerals brought in from the mainland as well agricultural produce. Their signalling system was by the block

**The PS** *Duchess of Fife* was jointly owned by the London and South Western and London, Brighton and South Coast Railways and was used on the service from Portsmouth to Ryde. It was built in 1899 by Clydebank Engineering and Shipbuilding Company and was withdrawn in 1928. *Pamlin Prints*

system. The passenger rolling stock was a mixture of four- and six-wheeled carriages and eight and twelve-wheeled bogie carriages. There were also some electric third rail tram cars running on Ryde Pier. Motive for the IWR comprised seven 2-4-0 side tanks some of which dated back to 1864 and were the oldest locomotives on the Southern Railway. The IWCR provided some 2-4-0, side tanks a 4-4-0 side tank and some 0-6-0 side tanks. The latter having been acquired from the LBSCR. The FYNR provided a 0-6-0 saddle tank and a 0-6-0 side tank. The latter again having been acquired from the LBSCR. The FYNR might have been more important than it was had the South Western and Isle of Wight Junction Railway (which had obtained its Act in 1901 to build a railway from the LSWR's Lymington branch to join the former railway between Freshwater and Yarmouth) been built. But the powers to build it had lapsed in 1921. In 1908, Siemens Brothers had offered to survey the IWR with the intention of electrifying the passenger services, but the financial outlay required for electrification was beyond the Railway Company's resources.

Whilst the IWR and FYNR used the Westinghouse air brake the IWCR used the vacuum brake.

The IWR's works were at Ryde, whilst those of the IWCR and the FYNR were at Newport.

Regarding signalling, the IWR used signal boxes designed by and lever frames built by Saxby and Farmer, the IWCR used signal boxes designed by and lever frames built by the Railway Signalling Company and the FYNR's signal boxes were a mixture of its own

**Isle of** Wight Railway locomotive No W16 *Wroxall* built in 1872 was one of a class of seven 2-4-0Ts built by Beyer Peacock between 1864 and 1883. *Wroxall* survived in service until 1933 and was the last member of the class to be withdrawn. *Photomatic*

**The Brading** to Bembridge branch was the Isle of Wight Railway's only branch line. In this photograph taken at the Bembrdge terminus IWR locomotive No W13 *Ryde,* built in 1864, heads a branch train. Ryde was withdrawn in 1932 and preserved until 1940 when it was scrapped. *John Scott-Morgan collection*

design and those of the Railway Signalling Company and its lever frames were built by the Railway Signalling Company.

Only the Bere Alston to Callington section of the PDSWJR was worked by the company: the rest was worked by the LSWR. The principal places served on the section the company worked were the Bere Alston and Callington. Passenger traffic was principally local or those holidaymakers choosing to visit that part of Devonshire. Goods traffic was coal, industrial materials and agricultural produce. Signalling was lock and block for a single line railway. Passenger carriages were second hand bogie eight-wheelers. There was one 0-6-0 side tank and two 0-6-2 side tanks.

**The Isle** of Wight Central Railway also had four 2-4-0Ts of the same type as those of the Isle of Wight Railway and this photo shows locomotive No 7, later W7, built in 1882, on a Newport train passing a Ryde train in about 1910. *John Scott-Morgan collection*

The PDSWJR used the vacuum brake and motive power and rolling stock was maintained at Callington.

The LBR served the places in its name or rather Lynton and Lynmouth and Barnstaple. It was the SR's only narrow gauge railway being built to the gauge of 1ft 11½in, the same as several narrow gauge railways in Wales. Its principal passenger traffic was holidaymakers and local traffic. Goods traffic was coal, industrial materials and agricultural produce. Signalling was lock and block for a single line railway. Passenger rolling stock was eight-wheeled bogie carriages. There were four 2-6-2 side tanks and one 2-4-2 side tank. Whilst the 2-6-2s were built in England, the 2-4-2 was built in America and was the only American built locomotive on the SR at the time of the grouping.

The LBR used the vacuum brake and motive power and rolling stock was maintained at Pilton Road, Barnstaple.

In 1926, the SR purchased the Newhaven East Harbour. This included a private railway running along the shore past an old tide-mill to near the Buckle Inn at Bishopstone. The SR did not carry out any development there.

The SDR jointly owned with LMS served Bath, Bournemouth, Burnham on Sea, Wells, Bridgewater and Templecombe. Its principle passenger traffic was holidaymakers from the north of England to Bournemouth and local passenger traffic. Freight traffic included coal from the Somerset coal field, industrial materials and agricultural produce. Signalling was lock and block for both double and single track sections. Passenger rolling stock was a mixture of six-wheeled stock and eight-wheeled bogie stock. Motive power consisted of 2-4-0, 4-4-0, 0-6-0 and 2-8-0 tender locomotives, 0-4-0, 0-4-2, 2-4-0 and 0-6-0 saddle tanks and 0-4-4 side tanks.

**Here are** two of the Lynton and Barnstaple Railway's locomotives. On the right is 2-6-2T No 759 *Yeo* built by Manning Wardle of Leeds in 1898 and on the left 2-4-2T No 762 *Lyn* built by Baldwin in 1898 and which was the first American built locomotive on the SR. *John Scott-Morgan collection*

The SDR works were at Highbridge.

For braking the vacuum braking system was used.

Its signal boxes were of its own design, but Stevens and Sons lever frames were used.

There were also what may be described as dependent companies. These were: the Kent and East Sussex running from Headcorn on the South Eastern and Chatham main line from London via Ashford to Folkestone and Dover, via Tenterden to Robertsbridge on the same Railway's line from London via Tonbridge to Hastings; the East Kent, running to Wingham Colliery from Shepherd's Well on the South and Chatham main line from London via Canterbury to Dover; the West Sussex or Hundred of Manhood and Selsey Tramway running to Selsey from Chichester on the London, Brighton and South Coast main line from London to Portsmouth; and the narrow gauge Rye and Camber running to Camber Sands from Rye on the South Eastern and Chatham line from Ashford to Hastings.

**THE MANAGEMENT OF THE NEW COMPANY**

The Chairman of the new company was Brigadier-General Sir Hugh Drummond, formerly of the LSWR. There were two Deputy Chairmen. Gerald Loder, formerly Chairman of the LBSCR and Brigadier-General the Hon Everard Baring of the SER and the SECR. Sadly Drummond did not last long in office, dying on 1 August 1924 and was succeeded by Everard Baring

The decision as to who would be the General Manager took slightly longer. There were three contenders: Sir Herbert Walker of the LSWR; Sir William Forbes of the LBSCR; and Percy Tempest of the SECR.

Having three General Managers to run the company was not ideal. It was almost as if the formation of the SR had not taken place. According to Michael Bonavia in *The History of the Southern Railway* (Unwin Hyman 1987) the Board of the company had given the three General Managers a remit to prepare an organisation for the Southern Railway as whole, but no progress could be made whilst they remained virtually independent of each other. Given that Drummond had been the Chairman of the LSWR, to explain 'that the weak-kneed Board decision to appoint three General Managers with equal status and duties was not working, his position was impossible.' Bonavia says that Drummond and the Board of the SR had leant over backwards so as not to give the impression that LSWR element of the company intended to dominate it. He says in not so many words that perhaps it would have been better to sort out who would be the General Manager for the whole company at the outset. Walker saw Drummond and offered to resign if that would make things easier, but Drummond refused and decided finally to deal with the situation. Having talked to the Board, it was arranged for Sir William Forbes to retire on 30 June with a 'golden handshake' or the 'offer that he could not refuse'. He was 67 years old. Next Percy Tempest was persuaded to retire on 31 December. To soften the blow he received a knighthood in the 1923 King's Birthday Honours and for one year after his retirement was appointed as a consultant to the SR. He died on 24 November 1924. This left Walker in sole command. According to the *Daily News* of 2 January 1924:

New General Manager of Southern Railway. The directors of the Southern Railway have appointed Sir Herbert Walker to be general manager. Sir Percy Tempest formerly general manager of the South Eastern line, has retired, and Sir Herbert Walker becomes supreme controller of the old South Western, Brighton and South Eastern lines. Sir Herbert entered the service of the London and North Western Railway as a clerk when 17 years of age. He climbed from post to post until in 26 years he became assistant to the general manager. In 1912 he was offered and accepted the office of general manager of the London and South Western Railway. He is a quiet, unassuming man of 55 years.

It would I think be true to say that Sir Herbert Walker was something of a forceful man. He was born in 1868, the son of a doctor, and had been intended for the medical profession, but for family reasons he had abandoned his medical studies and instead joined the LNWR, where he rose to the rank of outdoor goods manager. Charles Klapper in *Sir Herbert Walker's Southern Railway* (Ian Allan 1973) says that he carried out duties which would have rated the designation assistant general manager. But clearly not on the LNWR. At the end of 1911 he was feeling somewhat resentful at the lack of recognition of the special tasks he had performed on the LNWR, when he was head hunted by the LSWR for the post of general manager. Naturally he took it, commencing work on 1 January 1912. In 1914, at the beginning of the First World War, he was appointed Chairman of the Railway Executive Committee and in 1917 he was knighted. If he had a personal failing it was a lack of small talk – the shy restraint that held him back from conviviality. He did however have a very good memory for names and facts and figures and this stood him in good stead when meeting members of staff of all grades.

What of the other senior managers?

The Chief Operating Superintendent was E.C. Cox, whilst the Chief Mechanical Engineer was R.E.L Maunsell and the Locomotive Running Superintendent was A.D. Jones. All three came from the SECR

The Chief Engineer was A.W. Szlumper and the Chief Electrical Engineer was Herbert Jones, both of whom came from the LSWR. The Deputy Chief Engineer was George Ellson, who came from the SECR.

The Chief Commercial Manager was H.A. Sire and the Chief Accountant was Charles A. de Pury, both of whom came from the LBSCR.

The Company Secretary was John Jennings Brewer who had been the Secretary of the LBSCR. Unfortunately, Brewer died on 30 April after an operation. He was succeeded by Godfrey Knight and the Solicitor was William Bishop. Both came from the LSWR.

The headquarters of the new company was at Waterloo station in London except for the Docks and Marine department, whose headquarters was at Southampton.

In order to create a new unified SR image, in 1924 in an official statement the staff were asked to refrain from referring to the South Western, Brighton and South Eastern sections and instead refer to the Western, Central and Eastern sections. At Victoria, where the LBSCR and the SECR had their west end London terminus stations built next to each other, an opening was constructed in 1925 thus rendering the two stations one.

## ELECTRIFICATION

As indicated, the SR inherited three different electrical systems of which only two were in operation.

Whilst the LSWR had completed the first stage of its third rail electrification plan, the LBSCR had started work on extensions of its overhead electrification from Balham to Coulsdon North and Sutton.

At the beginning of its life, the SR was faced with a problem. According to John Glover in *Southern Electric* (Ian Allan 2001), from 1913 to 1923 the number of passengers carried by the three constituent companies of the SR rose by little more than a quarter, but because of the shortening of working hours after the First World War, morning and evening peak traffic had to be carried in one and a half hours less than in 1913. Under those conditions, operation with steam traction was becoming increasingly difficult. There was considerable congestion both in the London termini and their approaches. Line widening was out of the question owing both to cost and the time needed for construction. Whilst it is not to say that the operation of suburban services with steam could not be improved and the Great Eastern Railway (GER) which had the heaviest suburban traffic out of London had done that on its suburban service out of Liverpool Street with the 'Jazz Service' started in 1920, for the SR, electrification was the only realistic solution and therefore to press on with the schemes that had already been developed or were proposed. Even the GER admitted in 1921 that it had a scheme for the electrification of its suburban system. However, that took somewhat longer to realise and happened long after that company's demise..

Before going ahead with electrification, it was important to decide which system to use. Of the three possible systems to use, that of the LSWR was the most extensive, whilst it could be argued that that of the LBSCR was the most advanced system. No practical work had been done on that of the SECR.

As mentioned, at the time of the formation of the SR the LSWR's electrification extended from Waterloo to Wimbledon via Clapham Junction and East Putney, the Kingston Roundabout and the Shepperton branch, the Hounslow Loop, New Malden to Surbiton and Hampton Court and Surbiton to Claygate. The service to the latter had been withdrawn in 1919. It was the company's intention to electrify from Hampton Court Junction to Guildford via Woking and via Cobham and from Raynes Park to Effingham Junction via Epsom. It had plans to electrify to Ascot, Windsor and Weybridge as well to Farnham and an alternative route to Guildford. Only passenger services were and would be worked. The system used was the 600 volts dc third rail system

The LBSCR's electrification extended from Victoria to London Bridge via the South London line, Victoria to Crystal Palace (Low Level) via Streatham Hill and from Peckham Rye via Tulse Hill and Streatham Hill and from Crystal Palace to Norwood Junction and Selhurst depot. The company was in the process of electrifying to Coulsdon North via Balham, Selhurst and East Croydon, from Selhurst to Sutton via West Croydon and Tulse Hill to Streatham Common as well as intending to electrify to Wimbledon by two routes and to Epsom Downs. Approval in principle had been given by company's board to electrify to Brighton and Eastbourne. Only passenger trains were worked electrically. The system used was the 6600 volts ac overhead system.

The SECR had intentions to electrify all lines including branch lines in an area bounded by Gillingham, Tonbridge, Dorking and

Tattenham Corner. It was intended to work all trains within the area by electric traction. For through trains this would have meant a change of engine at the border stations. The system to be used would have been 1500 dc protected third rail.

In order to decide which electrical system to use, a Departmental Committee of chief officers was set up, presided over by E.C. Cox the Chief Operating Superintendent. Having taken advice from experts including Philip Gibb of the Pennsylvania Railroad of America, the Committee recommended the adoption of the 600 volts dc third rail system used by the London and South Western section, the deciding factor being the ease of installation and the low cost of construction and maintenance. Klapper says that the decision to use the third rail was justified by calculations made by H.F. Trewman in his book *Railway electrification; a complete survey of the economics of the different systems of railway electrification from the engineering and financial points of view*, published in 1924. According to him, for a 15-mile suburban service with stops 1 mile apart and a five minute headway, twenty-two three-coach trains would be needed. Rolling stock investment for dc traction would be £285,000 as opposed to £435,000 for ac traction. Although energy consumption would not differ greatly, ac trains would cost 1s 10¼ d per mile against 1s 5¾d per mile for dc trains. The SR's Board accepted the Committee's recommendation with the proviso that as the extensions to the LBSCR's 6,600 volts ac overhead were so far advanced, they should be completed and brought into service. According to Roger Kidner in *The Southern Railway*(Oakwood1958), the reason for the 'odd behaviour' was that the complaints about the service provided by the SR in the London area were becoming more and more bitter and had eventually reached 'political levels' and parliament was 'within an ace of taking drastic action'. He continues that there 'was no time for long-term plans for the dc electrification of the Brighton lines to mature, and the useless expenditure of £600,000 was a cheap price for keeping out of the Government's clutches.'

A question that one needs to ask is: Why did the LSWR decide on the third rail 600 volts dc system? The obvious answer is that the section from East Putney to Wimbledon would be used by its electric trains and those of the MDR and given that this was only a two-track section it would not have made sense to use any system other than that used by the latter company, albeit with current return by the running rails. The fact that an overhead system – be it dc or ac – was better was irrelevant in those circumstances.

When the Committee was set up and when it produced its recommendations is unclear. Certainly, from newspaper reports of late March 1923 it was reported that the contract for the conductor rails for the electrification of part of the South Eastern and Chatham section had been let. In May it was reported in the local press that work would soon be put in hand on the electrification of the section and that the system used would be the third rail direct current system in use on the London and South Western section. According to local press reports, work started in June 1923. At the same time, work was already in hand in erecting the gantries on the London, Brighton and South Coast section. It was reported that conversion to the third rail system would cost too much. It is not until November of 1923 that one reads that authorisation had been given to extend the London and South Western section electrification to Guildford and Dorking. Glover says authorisation was given on 6 December.

In electrifying the lines there was, particularly on the South Eastern and Chatham section, but not exclusively, quite a bit of incidental works that had to be done.

The new electric services came into operation as follows. Balham to Coulsdon North and Balham and Sutton via West Croydon on 1 April 1925 on the London, Brighton and South Coast ac overhead system. According to the *Daily News* of 1 April:

BOON TO SURREY RESIDENTS. ELECTRIC TRAINS TO SUTTON & COULSDON. New electric services from Victoria to Sutton and Coulsdon North are opened to the public this morning. A trial trip was made over the line yesterday by a train carrying directors of the Southern Railway and representatives of the Press. This is the first of the extensions of the electrification scheme, which will eventually consist of 630 miles of electrically operated track. The cost of the scheme will be nearly £8,000,000.

Victoria to Orpington via Penge East; Raynes Park to Dorking North via Leatherhead; Nunhead to Crystal Palace (High Level); Leatherhead to Effingham Junction; Holborn Viaduct to Orpington via Nunhead; and Claygate to Guildford via Cobham were on the third rail dc system by 12 July 1925. The *Hampshire Advertiser* of Friday 10 July said:

Southern Railway. CELEBRATION OF A GREAT ACHIEVEMENT. General the Hon. E. Baring. chairman of the Southern Railway speaking at Guildford on Thursday on the occasion of the first electric train to Guildford said that Sunday next would be a great day in the history of the Southern Railway for on that day they would open the first stage of the electrification of the South Eastern and Chatham section lines from Victoria, Holborn Viaduct, and St Paul's to Orpington, via Herne Hill, from St Paul's to Orpington and the Crystal Palace via Nunhead and also the South-Western section lines from Waterloo to Dorking, via Leatherhead, and to Guildford, via Cobham. On these and further stages of their electrical schemes they were committed to an expenditure of about. £5,500,000. That day they celebrated a great achievement.

On 21 September 1925, Hayes to Elmers End was electrified on the third rail dc system. This was to provide crew training.

On 28 February 1926, electric services on the third rail dc system commenced on the lines from Charing Cross and Cannon Street to Orpington, Bromley North and Addiscombe and Hayes.

The *Sevenoaks Chronicle and Kentish Advertiser* of Friday 26 February 1926 said:

The electrification of another large section of the Southern Railway has been completed, and the new augmented services will be begun early on Sunday morning. Electric trains will run from Charing Cross, Cannon-street, and London Bridge to Orpington, via Chislehurst, to Bromley, via Hither Green, and to Beckenham Junction, Hayes and Addiscombe, via Lewisham and Ladywell. The Hon. Everard Baring, presiding at the half-yearly meeting of the Southern Railway yesterday, referred to the great success of the electric train service, and indicated further suburban extensions to be carried out during the present year. It will, in fact, be possible before long to go to the Derby by electric train, as an extension to Tattenham Comer is contemplated.

On 19 July 1926 electric services began on the third rail dc system from Charing Cross and Cannon Street to Dartford via Greenwich, Blackheath, Bexleyheath and Sidcup. This was not the date of the first electric trains running to Dartford. A few had run during the General Strike of that year commencing on 12 May. At the end of the strike, the service had reverted to steam traction. On 6 June a restricted service of electric trains began running from Charing Cross and London Bridge to Dartford via Woolwich, Bexleyheath and Sidcup. In connection with the electrification, Cannon Street was closed from 28 June to 18 July. On 19 July the full electric service started from Charing Cross.

## THE THREAT OF COMPETITION FROM THE UNDERGROUND RAILWAYS

Coupled with electrification, there was the threat of competition from the Underground railways.

On 1 January 1925, the SR acquired the freehold of the ELR. Passenger services other than excursions were left in the hands of the MR, whilst goods and excursions were in the hands of the LNER, the latter two being steam operated. The Metropolitan, commencing on 1 January 1924, had taken over the maintenance of the signals, buildings and permanent way from the SR which had acquired the responsibility from the SER on 1 January 1923. The railway appeared in the SR passenger timetable. Station name boards on the ELR were green diamonds with the name on a bar across it as opposed to the red diamonds with the station name across on a bar used by the Metropolitan since 1915. In late 1925, Walker suggested to Robert Selbie, the General Manager of the Metropolitan that the latter Company should run trains off the East London on to the SR. The idea appealed to Selbie as the ELR was losing revenue to bus competition and he considered that the only salvation for it lay in the development of longer distance through services. The Metropolitan made some very careful studies into the possibility of running four six-car trains at peak hours and three at other times (Mondays to Saturdays only) between Hammersmith and Addiscombe. The proposal was to use existing Metropolitan electric stock adapted to run on both the third and fourth rail system of the parent company and the third rail system of the SR. Apparently some trial runs were made. Unfortunately, the motor coaches of both the Metropolitan and the Hammersmith and City Joint Committee stock which worked the existing Hammersmith to the ELR service were not man enough to enable the projected service to maintain their projected paths amongst the higher powered SR emus. Alan A Jackson in *London's Metropolitan Railway* (David and Charles 1986) says that the Metropolitan's terms were thought too onerous by the SR, which closed the discussion in the latter part of 1926 and adds that there was no change in 1929 when the Metropolitan raised the matter after such a service had been recommended by the London and Home Counties Traffic Advisory Committee. According to him, the SR steadfastly maintained that it would lose revenue if it paid the Metropolitan's proposed minimum charges. Whilst there is no indication of the reasoning in the SR's minutes, Jackson says that the truth may have been that the SR feared operating difficulties as its electric lines engendered more and more traffic. Klapper says that at the hearing of the SR's 1925 Bill, the London County Council tried to persuade parliament that there should be through services and through fares from ELR to SR stations, but neither the SR nor the Metropolitan were anxious to allocate capital to special new rolling stock. Apparently owing to loading gauge and sliding door difficulties, a run-on of SR electric rolling stock would not have been feasible. Whilst George Hally, the Metropolitan's Chief Mechanical Engineer, was experimenting with coaches incorporating four 275 hp motors which would have matched the SR's trains on station to station runs, the Company wanted to lay out its capital in the period 1927-31 on its own fast developing Extension Line traffics and so the proposal was dropped. Jackson says that the reason the trains with those motors were ordered was with the extension of the electrification from Rickmansworth to Amersham in mind, and presumably also to Chesham.

Apart from the Metropolitan and the East London, there were the twin threats of the Metropolitan District and the City and South London. For the sake of completeness I will take the story to its end, even though one of the lines in question was not built until after the the period of this chapter.

The threats came in two forms. With regard to the Metropolitan District it came in the form of the Wimbledon and Sutton Railway which had been promoted by local landowners, one of whom was a director of the MDR, for a railway between the two places via Merton and Morden in the hope of increasing the value of their land through housing development. It had obtained its Act in 1910. The intention was that the railway would be run as an extension of the MDR, whilst it was hoped that the residents of Merton and Morden would find the new railway useful it was also hoped that the residents of Sutton would also find the new railway useful as well as the service provided to it by the LBSCR was less than satisfactory. Problems were encountered in trying to raise the money for its construction and attempts to arouse interest from the Underground Group which owned the MDR or the LSWR, which could also have worked the railway, failed to produce anything useful. Following discussions with Albert Stanley, the Chairman of the Underground group in 1912 the Wimbledon and Sutton board was replaced by Underground group nominees and the railway became for all purposes an Underground group subsidiary. There were a number of changes proposed. The MDR felt that it could not provide a good service to Sutton if it had to share tracks between East Putney and Wimbledon with the LSWR and in 1912 it obtained powers to four track the section south of East Putney and in 1913 between the south end of Waltham Green (now Fulham Broadway) Station, and the viaduct section of the Parsons Green sidings, whilst Parsons Green Station would have been rebuilt with four platforms. The MDR agreed to pay interest on the capital raised by the LSWR which was to carry out the former. In 1914 the First World War broke out.

The threat from the City and South London Railway (CSLR) came in the form of an extension from Clapham Common to Morden.

The CSLR was the oldest electric railway in London and had opened in 1890. On 1 January 1913 it had been acquired by the Underground and some pretty drastic work had to be done to it as it had its own current collection system and its trains were hauled by small four-wheeled electric locomotives as opposed to multiple units. Between 8 August and 1 December 1924, the railway was reconstructed and at its northern end extended from Euston to Camden Town to join the Charing Cross, Euston and Hampstead Railway (CCEHR). In 1912, the Underground group had taken over the as yet unbuilt Edgware and Hampstead Railway. In 1922, the planners in the Underground group saw that if the green fields of Middlesex offered potential traffic, why not those of Surrey? In November 1922, a bill was put forward to extend the CSLR to

Morden where it would have a junction with the Wimbledon and Sutton and also to extend the CCEHR from Charing Cross to join the CSLR at Kennington. Walker went to parliament to defend his company against what was seen as a wedge into its territory. He said that whilst the SR would not object if the extension terminated at Tooting, anything else would result in all out competition. Some horse trading went on. It was agreed that the Morden extension of the CSLR would not join the Wimbledon and Sutton and the SR would take over the latter. An incidental outcome of the agreement was that the SR resumed services over the Tooting, Merton and Wimbledon line which had lost them as part of the war time closures on 31 December 1916. Services resumed on 27 August 1923. This also saw the return of LSWR section trains to Ludgate Hill station. The CSLR extension received its Act on 2 August 1923 and the SR arranged for the take over and winding up of the Wimbledon and Sutton. The extension to Morden opened on 13 September 1926. Other than at the terminus at Morden all of it was in tunnel. Work on constructing the Wimbledon and Sutton line began in October 1927. The first section to open was from Wimbledon to South Merton on 7 July 1929 followed by the rest of the line on 5 January 1930. The line was electrically worked from the start.

## THE ISLE OF WIGHT

The three railway companies on the Isle of Wight were something of a mixed bag. The Isle of Wight Railway was profitable, but it had the oldest locomotives on the SR – three dating from 1864. They were all 2-4-0 side tanks. The youngest was built in 1883. The Isle of Wight Central Railway was a rather shakier concern and had the most variety of locomotives. Some 2-4-0 side tanks built between 1876 and 1898, a 4-4-0 side tank built in 1890 and some former London, Brighton and South Coast 0-6-0 side tanks of the A1 and A1X Classes – better known as Terriers and built between 1872 and 1880. The Freshwater, Yarmouth and Newport Railway was in a very unhealthy state and was in receivership. It had two locomotives – a 0-6-0 saddle tank dating from 1902, the newest locomotive on the island, and a former LBSCR 0-6-0 side tank of the A1 Class built in 1876. The carriage stock was in a pretty bad state – mostly four-wheeled carriages. Most of it was second hand from the mainland. There was one striking exception – a former Midland twelve-wheeled clerestory carriage used for push-pull working by the IWC. In addition, the FYNR had a four-wheel petrol driven rail motor dating from 1913. The goods rolling stock was not in a good state either.

In order for the SR's Directors to gain an insight into the island's problems and possibilities, Walker arranged for them to visit it in August 1923. A result of the visit was the purchase by the Southern of Ryde Pier with its electric tramway. In 1927, the electric trams were replaced by petrol engine trams. They were numbered 1 to 4. Another thing that happened was the transfer of some LSWR O2 Class 0-4-4 side tanks to the island. By the end of 1926, twelve members of the class had been transferred. At the same time, four IWCR and three IWR locomotives were withdrawn. During the life of the Southern, all the locomotives of the constituent companies, other than members of the former LBSCR A1X 0-6-0 side tanks, were withdrawn, although one returned to the mainland before the end of the company's existence. A further three members of the class were sent to the island during the company's lifetime, as were four members of the former LBSCR E1 0-6-0 side tanks and a further nine O2s with another two following in 1949. Between 1946 and 1949, a former LBSCR E4 Class 0-6-2 side tank was tried on the island. Replacement passenger and goods rolling stock was sent over from the mainland.

In the summer of 1924, a passing loop was installed at Wroxall on the Ryde to Ventnor line and in conjunction with this, a new platform and overbridge were also constructed. In 1925, new structural improvement work was carried out at both Sandown and Shanklin stations. From Ryde St John's Road to Smallbrook Junction, the IWR

**Commencing in** 1923 a total of 21 former LSWR O2 0-4-4Ts were sent by the SR to the Isle of Wight. In this photograph No W19, formerly No 206, is seen on a train on the island. This locomotive and No W20/211 were transferred to the island in 1923. No W19 was later named *Osborne. John Scott-Morgan collection*

and the IWCR had their own separated single lines parallel to each other, but for the summer service of 1926 they were converted to a double line controlled by a new signal box. This arrangement was only in use during the summer and in winter single line working on each track into Ryde St John's Road was resumed. Also in 1926, the passing loops at Ashey and Whippingham on the Smallbrook Junction to Newport section were supressed and a new one installed at Havenstreet.

## THE LYNTON AND BARNSTAPLE RAILWAY

On the narrow gauge Lynton and Barnstaple, a number of improvements were made including enlarging the station buildings at Lynton, building a staff canteen alongside the engine shed at Pilton Yard in Barnstaple and erecting a new waiting shelter at Parracombe

Halt. A new locomotive of the 2-6-2 side tank variety was built for the line. The third class compartments had upholstered cushions fitted to the wooden seats. In 1926, two new travelling cranes were bought from Messrs George Cohen and Co and the following year a match truck was built for one of the cranes by the SR.

## THE BASINGSTOKE AND ALTON LIGHT RAILWAY

The Basingstoke and Alton Light Railway had been built by the LSWR under a light railway order dated 9 December 1897 and had opened for traffic on 22 May 1901. The railway was just under 12½ miles long and ran between the two places in its name. The purpose of the railway seems to have been to prevent the GWR getting to Portsmouth. Another line built by the LSWR was the Meon Valley line running from Alton to Fareham and which opened on 1 June 1903.

Traffic on the Basingstoke and Alton was not particularly heavy and when in the First World War the government demanded that the railways supply 200 miles of permanent way material to strengthen the lines of communication in France, the light railway was an easy target for twelve miles of track required from the LSWR. Closure took place on 1 January 1917. After the end of the war, there were hopes in the area served by the Light Railway that it could be re-opened. One of the preparatory works that had been done by Walker as the LSWR's General Manager was to insert clauses for the abandonment of the Light Railway in that Company's Bill for 1922-3, which was automatically continued as a SR measure. Walker and the SR came unstuck. The Select Committee of the House of Lords struck out the abandonment clauses in the Bill as inimical to the interest of the district. In consequence, the SR's counsel had to promise to restore the track and work the line for at least ten years. The SR acquiesced to the Lords' demands and the track was reinstated within twelve months on 18 August 1924. There was a sting in this as, according to the *Hampshire Telegraph* of 22 August 1924 in its account of the line's reopening, the line was to be worked for a period of ten years and if it was still worked at a loss it was considered only just and reasonable that the Company then be entitled to reconsider the whole position – i.e. close it.

## CHANGES TO SERVICES AND THEIR CONSEQUENCES

The summer services in 1923 had been organised by the timetable departments of the constituent companies. In 1924, under the supervision of Edwin Cox the Chief Operating Superintendent, they were done for whole railway. Unfortunately some rationalisation took place and the whole of the fast London to Portsmouth services were based on the LSWR line, leaving the poor old LBSCR line with a somewhat pedestrian service. There was a justifiable outcry and the fast services to Portsmouth on the LBSCR line were reinstated. A similar thing happened with the Hastings and Bexhill service being concentrated on the SECR line with the LBSCR line again losing out. Again this was remedied. Walker did not publicly upbraid Cox but took the blame himself as he wanted Cox to use his know-how on getting the electrified suburban services right.

## DEPENDENT COMPANIES

March 1925 saw extensions of the East Kent Light Railway. The first was from Wingham Colliery to Canterbury Road Wingham, a distance of 1 mile . The station was basically a terminal halt on the north side of the said mentioned road. There was an intermediate station at Wingham Town. The railway had hopes of getting to Canterbury, but Wingham was the nearest that it got to it. A 2½ mile branch was opened to Richborough from Eastry on the line to Wingham., but despite a halt being built at Richborough because the bridge over the Stour was not suitable for passenger trains passenger services terminated at the previous station Sandwich Road.

## MOTIVE POWER

The SR's first Chief Mechanical Engineer was Richard Maunsell. He was an Irishman who had become Chief Mechanical Engineer of the

SECR in December 1913. He had previously occupied the same post on the Great Southern and Western Railway in Ireland. He was born in 1868 in Raheny, County Dublin.

In 1923 he was offered the post of Chief Mechanical Engineer of the SR following the rejection of the post by Robert Urie, the Chief Mechanical Engineer of the LSWR. The other candidate was Lawson Billinton, the Chief Mechanical Engineer of the LBSCR, but who had less experience than Maunsell as he had spent time in the British army during the First World War.

Maunsell was a good administrator. He had assembled at Ashford Works on the SECR a very able team of men from the Great Western, Midland and Great Southern and Western railways. He was a firm, but fair man, who would not broach any nonsenses or tomfoolery.

If Maunsell had a problem it was that Walker was very determined on electrification. Maunsell would have flourished on the LNER under Wedgewood in the same way that Gresley would have been stifled on the SR under Walker

Maunsell, like Gresley on the LNER was happy to build new locomotives of classes not designed by him and which had originated on one of the various constituent companies of the SR.

Prior to 1931, locomotives maintained the numbers given to them by their constituent companies, prefixed by a letter denoting the main works of that company – thus ex SECR locomotives had their numbers prefixed by A for Ashford, ex LBSCR locomotives by B for Brighton and ex LSWR locomotives by E for Eastleigh. New locomotives had their numbers prefixed by the works at which they were built. Isle of Wight locomotives had their numbers prefixed by W for Wight. All locomotives transferred to the Isle of Wight were given new numbers e.g. ex LSWR 0-4-4 side tanks Nos E205 and E215 transferred to the Island in 1924 were renumbered W19 and W20.

On the LSWR, the Chief Mechanical Engineer Robert Urie had built three very successful classes of 4-6-0 between 1914 and 1922, the mixed traffic Class H15 coming out in 1914, the express passenger

Class N15 coming out in 1918 and the goods Class S15 coming out in 1920. These locomotives had only two cylinders. Urie's predecessor on the LSWR was Dugald Drummond who had had built five classes of four cylinder 4-6-0 which were not a great success.

Class H15, which were mixed traffic, had first appeared in 1914 and prior to the grouping a total of ten plus one notionally rebuilt from one of Dugald Drummond's generally unsuccessful four cylinder 4-6-0s had entered traffic. To these were added in 1924-5 a further ten locomotives of which five were notionally rebuilt from Dugald Drummond's four cylinder 4-6-0s. Urie's Class N15 express passenger two cylinder 4-6-0s first appeared in 1918. Seventeen had been built before 1923 and three more came out that year. To these Maunsell added a further fifty-three of his own version in 1925-7. These were all later given names associated with the Round Table and were known as King Arthurs. Class S15 goods two cylinder 4-6-0s first appeared in 1920 and twenty were built before the grouping. To these Maunsell added a further twenty-five of his own version in 1927-36. Of all the SR's 4-6-0s they were probably the best.

On the SECR Maunsell had designed three classes and rebuilt three. These were respectively new classes Class N two cylinder 2-6-0 and Class K two-cylinder 2-6-4 side tank both of which first came out in 1917 and Class N1 three cylinder 2-6-0 which although designed in South Eastern and Chatham days did not appear until March 1923. Whilst twelve Ns had appeared before the grouping, only one K had.

The rebuilds: were Class S 0-6-0 saddle tank which consisted of one locomotive was a rebuild of a Wainwright Class C 0-6-0 tender locomotive dating from 1900 and was rebuilt in 1917. Class E1 4-4-0 which first appeared in 1920 were rebuilds of members of Wainwright's Class E of the same wheel arrangement which had first appeared in 1906, a total of eleven locomotives being rebuilt. Class D1 4-4-0, which first appeared in 1921, were rebuilds of members of Henry Wainwright's Class D of the same wheel arrangement which

**Maunsell was** quite happy to build his own versions of Urie's own 4-6-0s and between 1925 and 1927 built fifty-three of his version of Urie's N15. All the N15s were given names of people or places associated with the legend of King Arthur. Here is No 771 *Sir Sagramore. Photomatic*

first appeared in 1901. A total of twelve locomotives were rebuilt before the grouping. For the record, Maunsell designed a Class C1 0-6-0 which was to be a rebuild of Class C. This rebuild never took place.

Following the formation of the SR, Classes N, N1, K and D1 were multiplied. Of the former, a further sixty-six were built up to 1934. Some of the locomotive boilers had been built by outside locomotive builders and the other parts by Woolwich Arsenal to provide alternative employment there after the end of the war. In December 1924 Class N No A819 was fitted with a Worthington-Simpson No1 feed water heater and pump on the left hand running plate. Despite showing fuel savings, Maunsell felt that after taking into account the cost of purchase, installation and maintenance, the savings were insufficient to warrant the equipment being employed in normal service and it was removed in November 1928 after 78,219 miles. The only other N1s built was a batch of five in 1930. Class K was better known as the River Class and were named after rivers by the SR; a further nineteen were built in 1925-6. There was also a solitary Class K1 three-cylinder 2-6-4 side tank built in 1925.A further twenty Ks were partially constructed in 1927, before fate intervened in the shape of the Sevenoaks accident. A further nine Ds were rebuilt to D1 in 1926-7 and the rest of the class would have been rebuilt, but the company's Civil Engineer passed the King Arthur Class 4-6-0s for passage over the former London, Chatham and Dover main line to Dover.

Whilst no former LBSCR designed locomotives were built after the Grouping, ten of Robert Billinton's Class B Class 4-4-0s, first built in 1901, were rebuilt to Class B4X in 1923-4. These were in addition to two members of the class rebuilt in 1922 by his son Lawson Billinton. As well as the B4Xs all of Douglas Marsh's Class I1 4-4-2 side tanks dating from 1906 and which were bad steamers were rebuilt to class I1X either with new boilers or with boilers from those Class B4 4-4-0s which had been rebuilt to Class B4X or in one instance from a Class I3 4-4-2 side tank. This was as an alternative to the original idea of withdrawing them. The idea for the rebuilding came from the Brighton Works manager. In their rebuilt form they performed satisfactorily. Rebuilding took place between 1925 and 1932.

The SR's most prestigious services were the Continental boat trains, the service to Brighton and the service to the West Country. In many ways, the former was the most prestigious, particularly those of the SECR.

The LBSCR and the LSWR had large locomotives for their prestige routes, be they 4-6-0s on the LSWR and large tank locomotives and 4-4-2s on the LBSCR.

The problem for the SECR was that the Railway's Civil Engineer would not allow anything larger than a 4-4-0 on the lines to the Kent

Coast and Hastings. Maunsell's predecessor, Harry Wainwright had during his tenure of office from 1899 to 1913, designed both 4-4-2s and 4-6-0s, but which were rejected as was a large 4-4-0 in 1907. A new class of 4-4-0s was built in 1914 just around the time of the outbreak of the First World War – Class L. Whilst it could work over the South Eastern section of the Railway, it could not work over the London, Chatham and Dover section. In 1915, following representations from the Locomotive Committee, Maunsell prepared a design for an outside cylinder 4-6-0 whose maximum axle load was 17½ tons and weight was 63 tons. There was nothing wrong with the design and the Railway's Civil Engineer had no problems with it, although he did insist on track alterations at Chatham and Maidstone East and the strengthening of two under-bridges between Victoria and Herne Hill. This was not the right time to carry out those works. Coupled with this the Railway's Drawing Office was fully committed to two new classes – the N Class 2-6-0 and the K Class 2-6-4 side tank. The Drawing Office was also committed to providing drawings of Belgian locomotives evacuated to Northern France ahead of the German invasion in 1914. In March 1917, the Board of the SECR said that in future all boat trains to Dover and Folkestone would run from Victoria. The rebuilding of some members of classes E and D had been Maunsell's solution to the problem.

Following the creation of the SR, once the permanent way and certain under-bridges had been strengthened ten new King Arthurs were sent to the Eastern (former SECR) section and from the beginning of the summer timetable on 13 July 1925, took over the Victoria to Dover and Folkestone boat expresses from the E1 and D1 class locomotives. Initially they were restricted to services via Tonbridge, but in 1926 the alternative route via Swanley, Maidstone East and Ashford became available and in 1927 the class could also be used over the old LCDR main line through Faversham to the North Kent coastal resorts and Dover. The locomotives could not be used on the Tonbridge to Hastings line with its restricted width tunnels.

In 1926, fifteen new inside cylinder 4-4-0s of Class L1 were put into service. The reason for the class was that in June 1922 the SECR had put into service a number of 80-minute non-stop Pullman Car expresses on the SER Section between Charing Cross and Folkestone Central. Normally the load for the trains was 220 tons, but at busy times this was exceeded and Wainwright's Class L 4-4-0s had difficulty keeping time, particularly on the Down trains. A D1 Class locomotive No 749 was tried for a week in September 1922 and performed satisfactorily. Whilst the D1 and E1 locomotives were ideal for the service they could not be spared as they were the largest locomotives the Civil Engineer would allow on the LCDR Section and so for the time being the Class L locomotives had to soldier on until new locomotives could be designed and built. As a temporary measure after the formation of the SR, ten former LSWR Drummond Class L12 4-4-0s dating from 1904-5 were transferred to the Eastern section. What emerged from the drawing office was an inside cylinder 4-4-0 designated class L1. The new class was similar to classes D1 and E1, but the smokebox and chimney were the same as those used on Class N and the cab of the new class was given side windows. The class proved a great success, being used on the Charing Cross to Folkestone 80-minute service, the Charing Cross to Hastings expresses and the Ramsgate semi-fasts. *The Locomotive Magazine* for April 1926 carried a photograph of No A759 and reported in an article beginning:

FIFTEEN new 4-4-0 type express engines have been built by the North British Locomotive Co., L td ., for the Eastern section of the Southern Ry. In general design the engines are duplicate with the "L" Class introduced in 1914, but improvements have been made by Mr. R.E.L. Maunsell, the chief mechanical engineer, in various details.

The article then went into various technical details and concluded 'These engines, which are numbered A 753 to A 759 and A 782 to

A 789, are mainly intended for the Charing Cross, Folkestone and Deal, and the Charing Cross and Hastings services.'

The Lord Nelson Class four cylinder 4-6-0s came about from the SR's Traffic Manager wishing to operate 500 ton Continental expresses on the Eastern Section. Good as the King Arthur Class was, it did not have sufficient reserve power to cover delays en route and any time lost would create havoc with the electrified suburban service. A new and more powerful class was essential. As a preliminary measure, Maunsell approached his fellow Chief Mechanical Engineers of the other three companies to ascertain the maximum axle load of their modern classes. George Hughes on the LMS said 20 tons for three classes, Nigel Gresley on the LNER said 20¾ tons was the highest axle loading then in use and Charles Collect on the GWR reported that when engineering and bridge works on the main lines had been completed, the then current 20 tons would be raised by 10 per sent for four-cylinder locomotives. Still not convinced, Maunsell sought information from several private manufacturers and foreign railways to determine whether locomotives of the size and power demanded by the Traffic Department could be constructed within the weight specified by the Civil Engineer. Another thing to be decided was whether the new express class should be a 4-6-0 or a 4-6-2. Maunsell sent his assistant James Clayton on some footplate trips. The first was on one of Collect's four-cylinder Castle Class 4-6-0s from Paddington to Plymouth and the other was on one of Gresley's three-cylinder A1 Class 4-6-2s from King's Cross to Grantham. The Castle Class won hands down. A four cylinder 4 6 0 it was to be. In March 1925, Maunsell had drawings prepared for a four cylinder 4-6-0 having 21½ tons on the bogie and leading and centre coupled wheels. The Civil Engineer found this excessive and insisted on a reduction of five per cent. By modifying the drawings through tapering the boiler barrel and reducing its length by 10½ inches, fitting lighter motion and cutting additional lighting holes, Maunsell was able to achieve this. Initially only one locomotive was ordered in June 1925 from Eastleigh Works

– No E850 *Lord Nelson*. It was delivered on 12 August 1926 and did a couple of trial runs, but following adjustments to the locomotive and its tender it did not return to service until 21 September. Amongst new features for an Eastleigh built locomotive was a raised Belpaire firebox instead of a round topped one, a cab with side windows and a firebox in which the grate was in two sections – the rear horizontal and the front sloping. The drive was divided between the front two axles whilst a 135 degree crank gave eight beats per wheel revolution to gain more uniform torque and firebox draught than possible with quartered cranks.

*The Locomotive Magazine* for September 1926 reported 'Southern Ry.—Mr. Maunsell's four-cylinder express locomotive, No. E850 *"Lord Nelson,"* has been completed at Eastleigh, and is now running trial trips.' The same magazine for October carried a photograph of *Lord Nelson* and reported

SOUTHERN RAILWAY. – NEW EXPRESS LOCOMOTIVE, "LORD NELSON" CLASS. IN order to meet the increasing weight of fast passenger traffic on the Southern Ry., a new and heavier type of locomotive, built at the company's works at Eastleigh to the designs of Mr. R. E. L. Maunsell, the chief mechanical engineer, has just been introduced. This engine, named *"Lord Nelson"* and the first of a class which will be known as the "Nelson" class, is a 4-6-0 four-cylinder simple superheater engine with double bogie tender, slightly heavier and more powerful than the well-known "King Arthur" type. The drive of the four cylinders is divided between two axles and the angles of the cranks are so arranged as to give eight separate impulses per revolution of the wheels, with the object of obtaining more uniform "torque" and more regular firebox draught than is possible in a four-cylinder engine arranged with "quartered" cranks. When other engines of the "Nelson" class are built they will be named after the following Sea-Kings:

*"Lord St. Vincent," "Lord Howe," "Lord Rodney," "Lord Hood," "Lord Hawke," "Howard of Effingham," "Sir Francis Drake," "Sir Walter Raleigh," "Sir Richard Grenville," "Martin Frobisher."* It should be stated that the new class in no way renders the "King Arthur's" obsolete; on the contrary, a further series is about to be constructed. On Tuesday, the 12th inst., the 11-0 a.m. *"Atlantic Coast Express"* was worked by the *"Lord Nelson"* between Waterloo and Salisbury, returning on the corresponding up train due to leave Salisbury at 2-26 p.m. Leaving Salisbury ten minutes late, the train arrived at Waterloo only two minutes behind time. The sixty-nine miles from Grateley to Clapham Junction were covered in sixty minutes. Andover was passed at 74 m.p.h. and Woking at 80 m.p.h. Between Basingstoke and Hampton Court Junction the average speed was 70 m.p.h. A party of Press representatives, accompanied by the chief mechanical engineer, assistant superintendent of the line, and other officials of the Southern Ry., travelled by the train, and were well satisfied with the excellent performance of the new engine.'

The edition for November said:

After having been in service on the Waterloo- Bournemouth trains, as well as working the *"Atlantic Coast Express,"* the *"Lord Nelson"* has been tried on the Eastern Section of the Continental service, as the overall dimensions will permit of its working on the main lines of any section of the Southern Ry. system. It is now to be fitted with a shelter for indicating purposes, and is stationed at Nine Elms … A striking poster has been issued by the Southern Ry. with a large illustration in colours of their new four-cylinder express locomotive *"Lord Nelson."*

The principal particulars are followed by the announcement in bold type 'The most powerful passenger locomotive in Great Britain.

Its Tractive effort being 33,510 lbs.' *Lord Nelson*'s claim as the most powerful passenger locomotive in Great Britain was short lived as in June 1927 the GWR turned out the first of its King Class four cylinder 4-6-0s, which had a tractive effort of 40,285lbs.

According to J.E. Chacksfield in *Richard Maunsell. An Engineering Biography* (Oakwood 1998), a final test took place on 10 April 1927 when *Lord Nelson* hauled a special train of sixteen coaches weighing 521 tons from London to Salisbury. Coal consumption on the run was 66½ lbs per mile, almost double that which had been achieved on earlier trials, albeit with lighter loads. Although the high coal consumption indicated the potential problems with steaming that lay ahead and which were never satisfactorily resolved, Maunsell felt the overall performance was deemed acceptable for a further ten locomotives to be delivered in 1928-9. This was followed by a further order for five in 1929. The order should have been for a further ten, but it was reduced to five. The result was that instead of twenty-one Lord Nelsons, only sixteen were built. Sixteen was not sufficient for a class of locomotive which had a split-level firebox, unlike the rest of the SR's steam locomotives, to be used to maximum effect on all the services for which they were planned, particularly when two or three would be unavailable due to maintenance and overhauls. I have said earlier that Sir Herbert Walker was something of a forceful man. According to Chacksfield, he was a passionate believer in electrification and it appears that his 'tremendous personality' held such a sway such that even Maunsell was not able to provide more convincing arguments for more Lord Nelsons to be built and indicated a reticence or inability on his part to make the case for the adequate funding for this and thus for them to be used over a wider area.

One only has to look at the following statistics in relation to new express 4-6-0 locomotives built in the years after 1926. On the SR, sixteen Lord Nelsons were built between August 1926 and November 1929. On the GWR, thirty Kings were built between June 1927 and August 1930. On the LMS, seventy Royal Scots were built between

September 1927 and October 1930 plus one rebuilt in 1935 from an experimental high pressure locomotive. On the LNER, thirty-seven Sandringhams were built between December 1928 and July 1931. A further thirty-six were built between March 1933 and July 1937 making a total of seventy-three in all.

When handled correctly, the Lord Nelsons were first class locomotives, but there were just not enough of them for the engine crews to get used to them. Maunsell tried various modifications, but without success. One, No 859 *Lord Hood*, was built with smaller 6ft 3in driving wheels in place of the existing 6ft 7in ones to see if this would improve performance over the heavily graded London-Dover line, but the difference was marginal. No 860, *Lord Hawke*, was built with a longer, heavier boiler but again with little improvement and in 1933 No 865, *Sir John Hawkins*, had its cranks modified to the more conventional 90 degree setting whilst No 857, *Lord Howe*, was fitted with a nickel steel boiler embodying a combustion chamber and a round topped firebox. In 1934, No 862, *Lord Collingwood*, was fitted with a double chimney and Kylchap blast pipe. In 1931, consideration was given to converting No 859, *Lord Hood*, to a four-cylinder compound. There is a certain irony in that on the LMS the Chief Mechanical Engineer, Sir Henry Fowler, had designed a four-cylinder compound 4-6-2 and construction of the first locomotive had started at Crewe Works when James Anderson, the Superintendent of Motive Power, prevailed on the company's board to cancel the project. A Castle Class 4-6-0 was borrowed from the GWR and given trials on the LMS and the company tried to borrow a set of Castle class drawings from the GWR but was unable to do so. Fowler then asked Maunsell for a set of Lord Nelson Class drawings, which the latter was happy to provide. The Royal Scots, as the class came to be known, had the same split-level firebox as the Lord Nelsons and proved to be an immediate success.

It was the failure of the Southern Railway to build more that caused the problem. Under British Railways, thirty of Riddles' Standard Class 5s 4-6-0s and fifteen Standard Class 4 4-6-0s were allocated

**Maunsell's Lord** Nelson Class had the misfortune to be built in insufficient numbers for engine crews to become fully acquainted with the locomotives with their split level firebox. No E850 *Lord Nelson* is seen here shortly after construction and fitted with indicator shelters. *Photomatic*

to the Southern Region and these had split level fireboxes to which engine crews became accustomed. These locomotives hauled some of the last main line steam trains out of London before the Waterloo to Bournemouth and Weymouth line was electrified in July 1967. Now, if only Walker had not had such a strong belief in electrification.

## ELECTRIC MULTIPLE UNITS

Up to the first four-car suburban emus in 1941, all suburban emus were three-car. The oldest of these dated from the LSWR's initial electrification of 1915 with the last being constructed in 1931-2. Constructed is not quite the right word as, other than fifty built in 1925 in two batches of twenty-six and twenty-four, all used the bodies of existing pre-grouping carriages, suitably modified. These were all classified 3SUB. In peak hours and other peak times, units would be supplemented by the use of two-car trailers; thus 3SUB two-car trailer 3 SUB making an eight car train.

For the Coulsdon and Wallington electrification of 1925 using the overhead ac system, five car units were built. The units comprised a motor luggage van sandwiched between two pairs of driving trailer and trailer cars

## SHIPS

The following ships were built during the period of this chapter.

John I. Thornycroft and Company of Southampton built *PS Shanklin* (launched in 1924 and entered service on 3 October 1924) for the Isle of Wight services:

In 1924, William Denny and Brothers of Dumbarton built SS *Dinard* and SS *St Briac* for the services from Southampton to Le Harvre / the Channel Islands / St Malo., For the Dover to Calais services TSS *Isle of Thanet* was built in 1925 and for the Folkestone to Boulogne services, TSS *Maid of Kent* was built in 1925.

The following ships were cargo boats on services from Dover to Calais and Folkestone to Boulogne built by D W Henderson and Co, Glasgow

TSS *Tonbridge* was built in 1924
TSS *Minster* was built in 1924
TSS *Hythe* was built in 1925
TSS *Whitstable* was built in 1925
TSS *Maidstone* was built in 1926

For handling perishable traffic from French ports and the Channel Islands, D W Henderson and Co, Glasgow built

TSS Fratton was built in 1925
TSS Haslemere was built in 1925
TSS Ringwoodwas built in 1926

## OPENINGS AND CLOSURES

In the years 1923 to 1926 there were three openings and two closures. The two closures were the result of one of the openings.

Both the SER and the LCDR had opened lines to Ramsgate and Margate. The SER opened a line from Canterbury to Ramsgate (Town) on 13 April 1846 and from there to Margate Sands on 1 December 1846. The line from Ramsgate Town to Margate Sands was not a straight continuation of the line from Canterbury but involved a reversal at Ramsgate. On 5 October 1863, the LCDR opened a line from Herne Bay via Margate to Ramsgate Harbour. The original plan of the railway was to terminate at Margate, but before it was completed, the extension to Ramsgate via Broadstairs was authorised. The approach to Ramsgate Harbour was through a tunnel.

It would be true to say that the layout of the railways in Margate and Ramsgate was somewhat complicated. The LCDR had two stations in Margate – Margate West and Margate East, whilst the SER station was at right-angles to them. The two companies' stations in Ramsgate were some distance away from each other and before the First World War the SECR had planned rationalisation, but the war prevented it. The plan involved linking Margate West to Margate Sands. Two bridges were built before 1914 and new platforms were begun at Margate West. This would have enabled LCDR section trains to reach either Ramsgate Town or Ramsgate Harbour. The SR's solution was to build a new section of line 1½ miles long from the LCDR line to Ramsgate to the SER line to Ramsgate. The new line opened on Friday 2 July 1926. On the same day, a new station was opened in Ramsgate, and Margate Sands, Ramsgate Harbour and Ramsgate Town stations were closed as were the lines from Ramsgate Town to Margate Sands and the line through the tunnel to Ramsgate Harbour. At Margate, a new goods depot approached from Margate passenger, as Margate West was renamed, was built on part of the former South Eastern line to Margate. The *East Kent Times and Mail* carried an article headed 'OFFICIAL OPENING OF

THANET'S NEW RAILWAY'. The first sentence read 'On Friday a new epoch dawned in the history of the Isle of Thanet, when the new railway line between Broadstairs and Ramsgate was formally opened.' The official opening took place that day when a party of the Southern's directors and officers led by the then Chairman Everard Baring and Sir Herbert Walker travelled down by the 9.05 am train from Victoria to Margate and then combined civic receptions in Margate, Broadstairs and Ramsgate with inspections of the new works. On 19 July, an intermediate station at Dumpton Park between Broadstairs and Ramsgate was opened.

The other two openings up to the end of 1926 were both on the former LSWR section. The first was the 9½ mile branch line from Totton between Southampton and Brockenhurst to Fawley, which was opened on Monday 20 July 1925 under a Light Railway Order. The *Hampshire Advertiser* of Friday 24 July carried an article headed 'OPENING OF TOTTON–FAWLEY LINE'. According to the newspaper there was no official opening of any kind.

The other line opened was the 20½ mile line from Halwill Junction to Torrington which opened on Monday 27 July 1925. The railway was built under the North Devon and Cornwall Junction Light Railway Order and was technically independent of the SR, although worked by it. The *Western Morning News* of Tuesday 28 July carried an article headed 'NORTH DEVON'S NEW RAILWAY. OPENING OF THE TORRINGTON–HALWILL LINE'. The newspaper said that the line was opened without any ceremony. The newspaper also carried photographs of the first passengers on the new railway. The largest place served by the railway was Hatherleigh, but unfortunately the station was two miles from the town and buses had already started running there from Okehampton.

There was another closure, but it was the SR's only bus service that it operated on its own. This was the former LSWR route from Exeter to Chagford which ceased on 20 September 1924 following competition from rival bus operators.

## STATIONS OPENED AND CLOSED

The following stations were opened between 1 January 1923 and 31 December 1926 including reopenings. They are in date order:

Mountfield Halt 1923
Merton Abbey reopened 27 August 1923
Bentworth and Lasham reopened 18 August 1924
Cliddesden reopened 18 August 1924
Herriard reopened 18 August 1924
Nunhead resited 3 May 1925
Motspur Park 12 July 1925
Fawley 20 July 1925
Hythe in Hampshire 20 July 1925
Marchwood 20 July 1925
Dunbear Halt 27 July 1925
Hatherleigh 27 July 1925
Hole 27 July 1925
Meeth Halt 27 July 1925
Petrockstow 27 July 1925
Watergate Halt 27 July 1925
Yarde Halt 27 July 1925
Ramsgate Town resited 2 July 1926
Deptford reopened 19 July 1926.
Dumpton Park (Ceremonial) 2 July 1926, (Public) 19 July 1926
Maddaford Moor Halt 26 July 1926

The following stations were closed between 1 January 1923 and 31 December 1926. They are in date order:

Whipton Bridge Halt 1 January 1923
Loughborough Junction platforms on Cambria Road Spur closed 12 July 1925
Midhurst (London and South Western) 13 July 1925

Margate Sands 2 July 1926
Ramsgate Harbour 2 July 1926
Ash Green (Goods) 1 December 1926

## EXTERNAL COMPETITION

The SR had competition for traffic with the GWR between London and Reading, Weymouth, Exeter, Barnstaple, Plymouth and Bodmin, but there was also competition with the LMS for the Gravesend traffic, the latter company providing through fares from London via ferry boats from Tilbury.

The up and coming motor bus companies were starting to create competition, particularly with rural branch lines, but also for the long distance traffic. The *Travel By Road Guide* for July 1923 has motor coaches serving the following places on the SR system from London: Bognor, Brighton, Bournemouth. Canterbury, Deal, Eastbourne, Folkestone, Hastings, Herne Bay, Hythe, Littlehampton, Margate, Portsmouth, Ramsgate, St Leonards, Southsea, Worthing.

The early airlines were starting to produce competition for the continental traffic. The first London to Paris air route had been started on 25 August 1919 by Aircraft Transport and Travel Ltd. Other companies also sprang up and other routes followed. In the face of competition from foreign competitors, Imperial Airways Limited was formed on 31 March 1924 by an amalgamation of the four largest companies British Marine Air Navigation, Daimler Airways, Handley Page Transport and Instone Air Line. The company was based at Croydon Airport to the south of London and initially operated year-round services to Paris, Brussels, Cologne and Amsterdam and summer services to Basel and Zurich as well as a service from Southampton to Guernsey.

In 1926, Imperial Airways put into service the three engine Armstrong Whitworth Argosy. This aircraft was considered luxurious for the time. The first passenger carrying flight was from London to Paris on 16 July 1926.

## PUBLICITY

According to Sir John Elliot in *On and Off the Rails* (Allen and Unwin1982), during its change from steam to electric traction for its suburban lines, the SR came in for a lot of criticism in the press for the delays and cancellations that occurred whilst the works were in progress. Walker was very friendly with Lord Ashfield, the head of the Underground group. One day, whilst Walker and Ashfield were having lunch, the former asked the latter how it was that the Underground group got a good press, but the SR was heavily criticised. Ashfield's reply was that the Underground group always kept its passengers informed of what was going on and it set out to cultivate public opinion. Above all, the Underground made it clear to the public what they were doing and why they were doing it. In that way, the travelling public felt that they were part of the operation and understood when things went wrong. What Walker needed, Ashfield said, was someone he could trust to deal with the Press and that he had a young man in mind who understood Fleet Street and had 'been brought up in it', the young man in question being Elliot. Elliot was the son of Ralph David Blumenfeld, who was originally American but had become a naturalised Briton in 1907. He was the editor of the *Daily Express* and was a friend of Ashfield. His son John was at the time the assistant editor of the *London Evening Standard*. Both the *Daily Express* and the *Evening Standard* were owned by Lord Beaverbrook. In 1923, Beaverbrook suggested John Blumenfeld change his surname to Elliot, which was his second Christian name. The logic for this being that Beaverbrook foresaw another war with Germany.

Elliot implies in *On and Off the Rails* that he was dismissed as Deputy Editor of *Evening Standard* in 1924 for running a story of a young woman's murder in a hut on the sands at Eastbourne. Elliot thought that it should be put on an inside page. Beaverbrook ordered it to be put on the front page. This landed the newspaper and its editor, but not its proprietor in court for contempt of court with a fine of £10,000. Beaverbrook blamed Elliott for carelessness even though the latter was following the former's orders against his better judgement.

On a cold morning in early January 1925, Elliot went to Waterloo for an interview with Walker. It is quite clear that the interview was arranged by Ashfield. According to Elliot, Walker told him that Ashfield had said that he could manage the Press. Elliot replied that whilst he could not, he could remedy the shortcomings of the SR as far as Fleet Street were concerned and through the newspapers inform the SR's passengers what it was planning and why there were so many delays at that time. When Walker asked him what he proposed to do, Elliot replied that whatever he did must be backed by Walker's personal authority. When asked how he proposed to operate, Elliot said that he would open an office at Waterloo to receive reporters and editorial writers so that any statement that he made on Walker's behalf or any of his staff would be made as from Walker. He added that he could not see any hope of success unless they worked along those lines. He continued that as Walker's man he would talk directly with each Head of Department about the major points of the electrification programme for which they were responsible so that he could build up a picture for the public.

Walker accepted Elliot's proposal and then asked him what his salary should be. Elliot said £2,000 on the grounds that the SR would not get anyone with his experience for less, to which Walker replied that that was more than his Divisional Managers got, thought about it and then agreed. He said that Elliot would work for him for a year after which they would agree whether Elliot remained with the SR or leave. After that the two went to see the Chairman of the SR, Brigadier-General Everard Baring and Elliot was introduced to him. It is quite clear from the conversation that followed that Baring was in favour of Elliot. Next the question of Elliot's job title arose and it was decided that he would be called the Public Relations Assistant.

The *Whitstable Times and Herne Bay Herald* of Saturday 21 February 1925 said:

Mr. John Blumenfeld Elliot has been appointed to the newly-created position of assistant to Sir Herbert A. Walker, General Manager of the Southern 'Railway. Mr. Elliot, whose headquarters will be at Waterloo Station, will take charge of the publicity work, act as "public relationships" officer of the Southern Railway, and serve as liaison between the general manager and the numerous public bodies in the district served by the Southern Railway. Mr. Elliot is' a son of Mr. R. D.Blumenfeld the editor the "Dally Express." and is a journalist. He is twenty-six years old.

Elliot soon got to work visiting each of the principal departments of the SR, as he had decided with Walker's agreement not to open a Press Office until he knew something of the operating problems involved in the electrification scheme. According to him, in that way he built up a 'lively picture of what the old-fashioned railways south of the Thames were going to be like when Walker's £25,000,000 suburban electrification was completed'. In the interim prior to the opening of the Press Office, Elliot saw to it that reporters who enquired at Waterloo would have their questions dealt with at once and this created a better atmosphere with the Press. When Elliot felt ready, he suggested to Walker that that they should have an open day at Waterloo for the Press at which he would be introduced. The open day went well and after that the Press Office was officially opened.

Elliot began a campaign of producing stories for the Press of the SR's modernisation plans and making known what was actively in hand and his efforts soon produced fruit, although inevitably there was some criticism.

After Elliot had been on the SR for a year, Walker sent for him and asked if he wanted to stay with the Railway or go back to journalism.

Elliot realised that he had fallen in love with the Railway and that his future lay with it and he told Walker that he would like to stay. Walker said that he was pleased with him and felt that he had a future with the SR and added that if he took Elliot on, the appointment would be permanent. Walker then called Godfrey Knight and told him of Elliot's decision and asked that it be recorded formally and to arrange for a Board Minute to be drawn up appointing him Assistant for Public Relations and Advertising at an annual salary of £2,000. Walker added that he would personally advise the Chairman. According to Walker, Elliot's job description, to use a modern term, was responsibility for advertising of all kinds including the letting of commercial advertising spaces on the railway.

Things blossomed. Posters, articles, books. Even locomotives. The 4-6-0 locomotives of the N15 Class, both those built for the LSWR and the newly built locomotives, were given names associated with the legend of King Arthur and the Knights of the Round Table. The K and K1 Class 2-6-4 side tanks were given names after rivers, whilst the former LBSCR H1 and H2 Class 4-4-2 locomotives were given names associated with the South Coast e.g. *Beachy Head*. The first of the new four cylinder 4-6-0s was named *Lord Nelson*. Locomotives sent to the Isle of Wight and some of the existing ones there were given names connected with the Island. According to Elliot, the naming of the King Arthur Class ruffled a few feathers. No 741 was named *Joyous Gard*. ASLEF, the enginemen's union, objected to the name and rang up, I assume, the Public Relations Office and asked, 'What about the bleedin' driver?' Sir Charles Morgan who was a former Chief Engineer of the LBSCR and a director of the SR was upset by No 750 named *Morgan le Fay*. Fay, for those who do not know, means fairy.

Up to this time there was only one named train wholly on the SR. This was the *Southern Belle* Pullman train which ran between London and Brighton and which had first run in 1908. In 1926 the company's best train to the West of England was named the *Atlantic*

*Coast Express*. The newly named train commenced running on 19 July 1926. The train has been described as not so much a train as more a service, having portions for Plymouth, Bude, Padstow, Ilfracombe and Torrington. There was another train which ran over the SR. This was the *Sunny South Express* operated by the LMS and the SR which had commenced operation in 1905 and ran from Liverpool (Lime Street) to Brighton via Kensington (Addison Road). It was originally operated by the LNWR and the LBSCR. The train ceased running at the outbreak of the Second World War. There was also the unofficially named *City Limited* which ran between Brighton and London and was used by City businessmen living in Brighton and which Bradshaw's Railway Guide honours with the title. Its ancestry went back to the early 1840s when the railway was opened through from London to Brighton.

Coupled with publicity could be held livery and numbering. Olive green for passenger and mixed traffic locomotives and carriages and black for goods locomotives were adopted. For the company's ships the funnel colour was buff with a black top.

At the Centenary Celebrations of the Stockton and Darlington Railway held from 1 to 3 July 1925 only one SR locomotive took part in the cavalcade. This was Class N15 4-6-0 No.449 *Sir Torre* on a train of new corridor stock. Also exhibited at the celebrations were the Canterbury and Whitstable Railway 0-4-0 *Invicta* and a Bodmin and Wadebridge Railway four-wheel carriage.

## SOUTHAMPTON DOCKS

The idea of improvements to Southampton Docks went back to before the grouping. In 1912 the LSWR had built the Ocean Dock for the White Star Line and whatever comparable shipping lines could be brought in. On 6 March 1920, the Cunard Line ship the *Mauretania* sailed from Southampton for the first time. In its last days, the LSWR decided on building a floating dock for 60,000 ton ocean liners. The scheme was approved by the Company's board in May 1922. It was built by Sir W.G. Armstrong, Whitworth and Company Limited and opened by the Prince of Wales on 27 June 1924.

According to the *Daily Chronicle* of 28 June:

PRINCE WORKS SCORE OF LEVERS. Opening World's Biggest Floating Dock. In reply to the civic welcome (given by Southampton), the Prince recalled that he used to watch the shipping which entered and left the port of Southampton when he was a naval cadet at Osborne, and said that the new floating dock was definite proof of the progress of the port. The dock, which has an 18,800 tonnage, can hold the largest liners afloat, The Prince, who was the first to enter the dock, pulled the levers to sink the dock, thus allowing to enter the 81,000 tons of water necessary to submerge it. There are no fewer than 70 levers to manipulate the dock, and the Prince personally operated one section, a quarter of the whole. After the lunch on the *Aquitania*, the Prince embarked on the *Duchess of Fife*, and steamed through the dock, breaking a red, white and blue ribbon across the entrance, and so completing the ceremony.

## ACCIDENTS

There was only one major accident during the period to the end of 1926 and that was on 4 November 1926 when the 4.15 am Victoria to Yeovil milk train became divided 530 yards east of Bramshott Halt on the Weymouth and West of England main line between Fleet and Farnborough. The crew of the train, believing that the issue could quickly be resolved, failed to inform the signalman or protect the rear of the train. The train was run into by the 5.40 am passenger express train from Waterloo to Bournemouth and Weymouth. The speed of the impact was about 40mph. In the collision, the driver of the passenger train suffered serious burns and scalds and died in hospital on 8 November. The fireman of the passenger train was also injured.

## STRIKES

From 21 to 29 January 1924 there was a strike by the Associated Society of Locomotive Engineers and Firemen (ASLEF). From reading contemporary national newspapers it is difficult to judge what level of service was provided, but it seems to have been variable. Some places had a service of sorts and some places nothing. Equally services varied from day to day. Not all drivers and firemen belonged to ASLEF and some belonged to the National Union of Railwaymen (NUR) and whilst some of these did not strike, some did in solidarity with ASLEF and there were also men who were not members of a trade union.

In May 1926 there was the General Strike. The Strike lasted nine days, from 4 to 12 May 1926 and was called by the General Council of the Trades Union Congress (TUC) in an unsuccessful attempt to force the British government to act to prevent wage reductions and worsening conditions for 1.2 million locked-out coal miners. Some 1.7 million workers went out, especially in transport and heavy industry. The government was prepared and enlisted middle class volunteers to maintain essential services. There was little violence and the TUC gave up in defeat.

According Bonavia the total number of trains run were as follows:

| | |
|---|---|
| 5 May | 338 |
| 6 May | 515 |
| 7 May | 753 |
| 8 May | 868 |
| 9 May | 274 – this was a Sunday. 212 were passenger and 62 goods |
| 10 May | 1,069 |
| 11 May | 1,370 |
| 12 May | 1,636 |
| 13 May | 1,655 |

Those trains that were run were run by volunteers, but not all had proper experience in running trains. According to Julian Symonds in

'*The General Strike*' (Readers Union 1957) the British railway system's prestige suffered a body blow during the strike as its dependence on skilled workers in a time of emergency was clearly shown.

To give example how the strike was reported, the *Daily Mirror* of Friday, 7 May said that the SR reported a continued improvement in main line and suburban services. The same newspaper for 11 May said that SR railway services were greatly improved.

*The Kent Messenger* of 8 May reported:

Porters in Plus-fours! Dover is enjoying the spectacle of railway porters in plus fours and blazers following the drafting to the town of about 50 students from Cambridge University on Thursday to work for the Southern Railway Company. They are divided into two shifts and are working eight hours each. Their duties include loading and unloading cross Channel boats and acting as firemen on trains. On Thursday they dealt with two incoming and outgoing steamers. They are sleeping in railway carriages at Dover Marine Station.

*The Whitstable Times and Herne Bay Herald* of 8 May said:

TRAINS RUNNING AGAIN ON THE WHITSTABLE AND CANTERBURY RAILWAY. DADDY PEARSON, THE FAMOUS ENGINE DRIVER, WHO RETIRED LAST WEEK, RETURNS TO WORK. The railway strike will soon be broken. From all parts of the country comes news of improved train services. The service on the Whitstable and Canterbury line was resumed yesterday (Thursday), the first train being brought from Canterbury by(George) "Daddy" Pearson, who only retired from the service last week on reaching the age limit. It had already been decided to make a small presentation to Mr. Pearson in recognition of his long services as engine driver and Mr. Warwick Wright, of Whitstable, had agreed to act as hon. treasurer to the fund.

This latest act of loyalty to the public on the part of Mr. Pearson will induce many to give their mite to the fund and we hope the result of the appeal will be a really substantial present for one who has served the travelling public loyally over such a long period.

I have mentioned how some electric trains were run to Dartford during the strike before the official start of electric services.

There were dangers involved. According to the *Daily Mirror* of 13 May:

While doing volunteer duty as a Southern Railway passenger train guard, Charles Moon, a young engineer's draftsman, of Guildford, was electrocuted near Charing Cross station. It was stated at the inquest yesterday that he stepped out of the train on a live rail, and was dead when released by William Tickner, an engineer.

Seven persons were injured on Tuesday (11 May) when a Southern Railway steam train from Ashford to Victoria was in collision with the rear of an electric train at Brixton Station. Police Superintendent Clarke rendered first aid until the ambulance arrived and took the injured to hospital. Some were detained. The last coach of the electric train was badly damaged. Traffic was delayed as a result of the accident.

During the strike, the volunteer engine crews became very attached to their engines and frequently booked on early to ensure that they were suitably cleaned for the day's work. The volunteer crew of class L No A763 unofficially named the engine *Betty Baldwin* after Esther Louisa (Betty) Baldwin, the youngest daughter of the Prime Minister Stanley Baldwin. The name was artistically inscribed in gold leaf on the leading driving wheel splashers and remained so until the locomotive was repainted in May 1927.

Bonavia says that on 13 May, Walker announced that men who went on strike would be re-engaged 'without prejudice to any question that may arise of their having broken their contract of service with the Company'.

According Klapper, 12,000 out of a force of 73,000 remained at work.

Following the end of the strike the miners remained on strike until forced by their own economic needs to return to work. By the end of November most miners were back at work. One effect of the miners remaining on strike was that because of problems in obtaining coal there was a 36 per cent reduction in the advertised train service.

# TO THE COMPLETION OF THE SUBURBAN ELECTRIFICATION 1927 TO 1930

## CHANGES AT THE TOP

In June 1927, the Chief Civil Engineer Alfred Szlumper retired and he was succeeded by his deputy George Ellson.

**This is** Plymouth Friary Station in about 1928 with a variety of motive power on the trains including an N Class 2-6-0 on the right hand side. Until the coming of the West Country Class, the N Class was the most powerful class that could work west of Exeter. *Photomatic*

## CENTENARIES AND PRESERVATION

13 to 20 September 1930 saw the celebration of the centenary of the opening of the Liverpool and Manchester Railway in Liverpool. The SR was represented at the celebrations by Lord Nelson Class 4 6 0 No E850 *Lord Nelson* and a combined first class dining saloon and kitchen car and a third class open carriage.

Earlier, 3 May had seen the centenary of the opening of the oldest section of the SR – the Canterbury and Whitstable Railway. *The Daily News* of Thursday 1 May carried a picture of the opening and said that plans had been made to celebrate the centenary, but all that happened, according to the *Whitstable Times and Herne Bay Herald* of Saturday 10 May 1930, was a talk given by the Rev. R.B. Fellows at the Canterbury Museum on the evening of Monday 5 May, who in the course of his talk commented upon the fact that neither the Railway Company nor anyone else had felt it was desirable to celebrate the

**The Canterbury** and Whitstable line was the oldest section of the Southern Railway, dating back to 1830 and lost its passenger service on 31 December 1930 – Centenary year. Here is R Class 0-6-0T No A124 at the head of a train on the line. *H.C. Casserley*

great event. At the talk there was on view a collection of early prints, plans, photographs, original minute and copy letter books, etc., dealing with the railway.

By the middle of autumn that year, the SR had decided to withdraw the passenger service from 1 January 1931 on the grounds of falling traffic and understandably there was opposition to this, but to no avail, with the last trains running on Wednesday, 31 December 1930. *The Whitstable Times and Herne Bay Herald* of Saturday 3 January 1931 carried an article which began:

THE LAST TRAIN. WHITSTABLE-CANTERBURY LINE CLOSED FOR PASSENGER TRAFFIC. The last passenger train to Canterbury from Whitstable started from the Harbour at 9 p.m. Wednesday, and twenty minutes later was back at Canterbury West, which station it had left just over an hour before. The passengers who made the journey both ways included a number of former servants of the Company, who had worked the line in their day. Naturally there were regrets at the closing one of the oldest lines in the country. Driver Gibbs, Fireman Crabb and Guard Relf no doubt shed a silent tear as they passed over the six miles of sleepers for the last time on a passenger train. It was naturally a connection which they felt hard to sever, but there was no alternative on the part the Southern Railway Company.'

The line remained open for goods traffic.

In either late 1926 or early 1927, the Stephenson Locomotive Society decided to purchase and preserve former LBSCR Class B1 0-4-2 No B618 formerly No 214 *Gladstone*. The Society approached the SR and the directors agreed to sell it for £140 to cover the cost of restoring the locomotive to as near its original state as possible. *Gladstone* was withdrawn from service in April 1927 and exhibited at Brighton station from 2 to 7 May and at Waterloo station with 4-6-0

No E850 *Lord Nelson* on 14 May. The intention was to put *Gladstone* on permanent display at the Science Museum in London, but renovations made this impractical and so it was exhibited at what was thought to be its temporary residence, but which became its permanent residence, the LNER's Railway Museum at York. Until the new build *Beachy Head,* this was the only example of a representative LBSCR express locomotive.

The *Worthing Gazette* of Wednesday 4 May said:

ON VIEW AT BRIGHTON THIS WEEK. Arrangements have been made between the Stephenson Locomotive Society and the General Manager of the Southern Railway (Sir Herbert Walker) for Mr. William Stroudley's engine, *"Gladstone"* — now ready for its journey to the Museum at York, where it is to be housed until accommodation is available for its reception at the South Kensington Museum, to be on view at Brighton Railway Station every day this week between 9 a.m. and 8 p.m. This historic engine has been restored to its original condition in every detail, and it is interesting to note that the man who painted it when it was first constructed, Mr. William Crick, has recently been repainting it in the distinctive colourings of the old London Brighton and South Coast Railway of the time. When first built. the "Gladstone" was regarded as the most powerful engine in England, and in that respect it was on a par with the Southern Railway's famous *"Lord Nelson"* of to-day.

*Reynolds Newspaper* of Sunday 15 May said, 'The Old Engine and the New.—Withdrawn from service, the locomotive *Gladstone,* built in 1882 for the London, Brighton, and South Coast Railway, was yesterday hauled into Waterloo by the *Lord Nelson* Britain's s powerful passenger engine. Over 2,000 people paid to see the engines.'

## ELECTRIFICATION AND A CLOSURE

25 March 1928 saw electric trains reach Crystal Palace (Low Level) from London Bridge and from Charing Cross to Caterham and to Tadworth and Tattenham Corner. The latter station that had been closed in September 1914 except for race traffic was now fully reopened.

Conversion of the overhead electrification now started. At the annual meeting of the SR held on 27 February 1925, shortly before the final extension of the ac overhead system, the Chairman of the Company in a statement said that one electrification system was enough and that the Brighton system was not compatible with that on the Western (LSWR) and Eastern (SECR) sections. He also made a remark to the effect that the problem was being studied by experts and that he was sure that they would find a satisfactory solution, even though it was fairly clear that the Southern would adopt the dc third rail system.

17 June 1928 saw the replacement of the ac overhead system with the dc third rail system from London Bridge to Victoria via the South London Line, the extension of electric services from London Bridge to Streatham Hill via Tulse Hill, London Bridge to Coulsdon North via Streatham and Streatham Common and via Norwood Junction, London Bridge to London Bridge via Norwood Junction and Selhurst, London Bridge and Epsom Downs via Norwood Junction and Selhurst and another route from London Bridge to Crystal Palace (Low Level).

1929 saw more electric services started. On 3 March, electric trains began running from London Bridge to Dorking North and Effingham Junction via Tulse Hill and Mitcham Junction and from Victoria to Beckenham Junction via Mitcham Junction. The same day, electric trains began running between Victoria and Beckenham Junction via Crystal Palace (Low Level), which replaced trains running on the overhead ac system between Victoria and Crystal Palace (Low Level). The commencement of electric services that day from Victoria and

Holborn Viaduct to Wimbledon via Tulse Hill and Haydons Road saw the cessation of services between Tooting Junction and Merton Park via Merton Abbey. As mentioned in the previous chapter, this service had been resumed on 27 August 1923 following its closure as part of the war time closures on 31 December 1916. The same day also saw the closure of Ludgate Hill station in London. 7 July saw the opening of the first part of the Wimbledon and Sutton line between Wimbledon and South Merton. The passenger services were electrically worked. The end of the overhead ac system came on 22 September when dc electric services began between Victoria and Coulsdon North and Sutton. There had been a brief overlap when it had been possible to see trains using the different systems together. The last ac electric train to run in service was the 12 10 am service from Victoria to Coulsdon North on Sunday 22 September. For a few short months, the overhead wires were kept live between Battersea Park and Peckham Rye to allow the ac stock to be worked to Peckham Rye car sheds for stripping of their electrical equipment before going to the works for conversion to dc working. Whilst the overhead wires were soon taken down, the supporting gantries were not removed until 1930 to 1932 with a few kept for signal gantries.

The *Portsmouth Evening News* of Friday 20 September 1929 said:

The Southern Railway announce that whole of their suburban electrified system- which was completed in the main last March, will, on and from Sunday-next, be run on the third-rail principle throughout. Most of the old Brighton section overhead track was converted prior to March, but owing to the great demands on rolling stock it was not possible to equip the whole of the routes with third-rail stock, .The difficulty has now been overcome, and the last train fitted with overhead equipment will run tomorrow. This will mean 'the final of the "alternating "' system of electrification, which was begun by the old London, Brighton, and South Coast Company in 1909, the original portion electrified by them being what is known as the South London Line, which incorporates the service from London Bridge to Victoria. The last stages of the overhead electrification were those from Victoria to Coulsdon North and Sutton, and curiously enough these two sections are the last to be transferred to the third rail principle

The *Daily Chronicle* of Friday 20 September 1929 said, 'On and from Sunday the third rail will replace overhead wires on all the SR suburban lines.'

The *Daily News* of Friday 20 September 1929 said:

LAST OF TROLLEY TRAINS. The whole of the Southern Railway's suburban electrified system has now been converted from overhead to the third-rail principle and the last train fitted with overhead equipment will run tomorrow. This change will mean the disappearance of the "alternating" system of electrification, which was begun by the old London. Brighton and South Coast Company twenty years ago.

Overhead electrification did not return to London until the commencement of the Liverpool Street to Shenfield electrification on the former GER on 26 September 1949.

1930 saw the completion of the Wimbledon and Sutton line when the section from South Merton to Sutton via St Helier was opened on 5 January. Again passenger services were electrically operated.

The suburban electrification could be held to have been completed on 6 July when electric trains commenced running from Whitton Junction and Hounslow Junction to Windsor, from Dartford to Gravesend Central and from Wimbledon to West Croydon via Mitcham. The extension to Gravesend Central gave the inhabitants of the town an all-electric route to London in addition to the all-steam routes to Gravesend West operated by the SR and the train and ferry route via Tilbury to the Town Pier operated by the LMS.

*The News Chronicle* of 7 July carried an article which read:

ELECTRIFYING THE SUBURBS. 50 MORE MILES OF LIVE-RAILWAY. A further 50 miles of "live-rail" way were brought into operation on the Southern electric suburban system yesterday, when three parts of line were changed over from the old steam system bringing the Southern's electrified tracks up to a total of 800 miles. The new electrified extensions are: Hounslow to Windsor, Dartford to Gravesend. Wimbledon to West Croydon. Windsor was 67 minutes from London by steam train. It will now be 46 minutes. Trains will run every 30 minutes during the busiest periods.

## OPENINGS AND CLOSURES

Apart from the Wimbledon and Sutton line, there were some other openings during the years 1927 to 1930.

At Minster, a short section of line known as the Minster Loop was opened on 7 July 1929.

The next opening needs a little bit of history. On 18 September 1871, the LCDR had opened a branch line from Nunhead to Blackheath. On 1 October 1888, the line was extended to Greenwich Park. The line had closed on 1 January 1917 as part of the SECR's response to a government request to provide extra capacity for war traffic.

Whilst most of the SR's freight traffic could be run at night to avoid impeding the frequent electric trains during the day, that which was exchanged with the northern companies had to be run in the day between the electric services. Trains to and from the former Midland and Great Northern lines and Hither Green marshalling yard ran via the City Widened Lines, the Snow Hill tunnel, Blackfriars Junction, Metropolitan Junction, London Bridge station and New Cross. The section of line between Metropolitan and Borough Market Junctions west of London Bridge was a particular source of congestion, as there was only one Up and one Down line available to carry all of the electric services into and out of Charing Cross station and the forty or so daily freight movements. The solution decided on in 1926 was to build two loop lines. One was from the site of the closed Lewisham Road Station on the derelict Greenwich Park branch; the bridge across the SER section line's St John's Station was removed and a new viaduct route involving a massive double span lattice-girder bridge was taken to come down alongside the North Kent line, with a double junction to give access to and from all four tracks in Lewisham Station.

The second loop was from the Mid-Kent line about 450 yards south of the country end of Lewisham Station and back to the slow roads of the Tonbridge main line on the way to Hither Green near Park Bridge Junction. This enabled the freight trains to travel via Loughborough Junction and Nunhead and only to appear on the main line after the North Kent and Bexleyheath trains had diverged and the lines were less occupied. The two loops opened on 7 July 1929 – but only for goods traffic. Chapman Frederick Dendy Marshall in his history of the Southern Railway published in 1936 calls the sections Nunhead to Lewisham and Lewisham to Hither Green. Passenger traffic began on 2 July 1930 but was only used by steam hauled excursions calling at Lewisham. Hither Green marshalling yard was also used by trains from the former GER line which travelled via Liverpool Street Station and the East London line and by trains from the former LNWR line which travelled via the West London, West London Extension and South London lines

Away from new lines, in 1927 the section of the former LCDR main line from Victoria between Kent House and Beckenham Junction was quadrupled, whilst on the former IWR the section between Brading and Sandown was doubled.

At Epsom, a new station was built or rather the former LSWR Station was rebuilt and the former LBSCR Station closed and trains diverted into the newly rebuilt station which came into operation on 3 March 1929.

At Dover, following a decision taken in 1924, the Southern finished the building of Dover Marine station which had been opened for the military by the SECR on 2 January 1915 and for the public on 18 January 1919. With the work complete, Dover Harbour station closed on 10 July 1927.

At Hastings, work was begun on rebuilding the station in 1930, but was not complete until 1931.

The SR suffered a slight loss of traffic at Folkestone in 1927 following a decision in 1926 by the Zeeland Steamship Company to change its English port to Harwich Parkeston Quay. The service to and from Folkestone ceased on 31 December 1926 and started from Harwich on 1 January 1927.

**STATIONS OPENED AND CLOSED**
The following stations were opened between 1 January 1927 and 31 December 1930:

Kemsley Halt 1 January 1927
Ashley Heath Halt 1 April 1927
Riddlesdown 5 June 1927
Sunnymeads 10 July 1927
West Weybridge 10 July 1927
Tattenham Corner reopened for all traffic 25 March 1928
South Bermondsey resited 17 June 1928
Aylesham Halt 1 July 1928
Farlington Halt 9 July 1928
Petts Wood 9 July 1928
South Merton 7 July 1929
Wimbledon Chase 7 July 1929
New Hythe 9 December 1929
Atlantic Park Hostel Halt 30 October 1929 –opened for special
    arrangements only
Morden South 5 January 1930
St Helier 5 January 1930

Sutton Common 5 January 1930
Sutton West 5 January 1930
Birkbeck 2 March 1930
Chestfield and Swalescliffe Halt 6 July 1930
North Sheen 6 July 1930
Swanscombe Halt resited 6 July 1930
Waddon Marsh Halt 6 July 1930
Whitton 6 July 1930

The following stations were closed between 1 January 1927 and 31 December 1930

Dover Harbour 10 July 1927
Leatherhead LSWR Section station 10 July 1927. All trains used the LBSCR Section station from 10 July 1927
Westcott Range Halt 1928
Mount Pleasant Road Halt 2 January 1928
Epsom LBSCR Section station 3 March 1929. All trains used the LSWR Section station from 3 March 1929
Ludgate Hill 3 March 1929
Merton Abbey 3 March 1929
Merton Park – Merton Abbey line platforms 3 March 1929
Browndown Halt 1 May 1930
Elmore Halt 1 May 1930
Fort Gomer Halt 1 May 1930
Drayton 1 June 1930

## THE THREAT TO CLOSE CHARING CROSS AND REPLACE IT WITH A STATION SOUTH OF THE RIVER

Since the formation of the London County Council (LCC) in 1889 there had been attempts to demolish Charing Cross Railway Bridge, close the station and replace it with a new station south of the Thames and replace the bridge with a road bridge. A leader of the movement was the radical politician John Burns. Burns' campaign carried on into the First World War before subsiding.

In the mid-1920s, it was discovered that John Rennie's Waterloo Bridge dating from 1817 had subsided and needed rebuilding. In 1925, a Royal Commission on Cross River Traffic in London considered various bridge schemes. The report of the Commission was published in 1926 and recommended that the existing Charing Cross Railway Bridge be replaced by a double–decked steel bridge with not more than five arches over the river providing for six railway tracks on the lower or present level with a 60ft roadway above and that two footways of 15ft each should be built immediately alongside the present railway bridge on its downstream side, with a new Charing Cross Station a little to the east of the existing one. The SR did not object to this in the secure knowledge that it would gain improved terminal facilities backed up by adequate compensation. After having consulted with the LCC, the Ministry of Transport requested the consulting engineers Messrs Mott, Hay and Anderson to investigate and cost the project in detail with the assistance of the LCC's chief engineer Sir George W. Humphrey and Alfred W. Szlumper who was the SR's chief engineer until his retirement in June 1927.

The engineers reported back in May 1928. Whilst they found the idea of a double deck bridge perfectly feasible, they proposed two alternative schemes. These were: a new Charing Cross Station (with roads each side) and a double-decked bridge, to cost £13,050,000; or a single road bridge, and the removal of the station to the south side of the Thames, to cost £10,770,000. Szlumper would not, naturally, endorse the second scheme. The government headed by Prime Minister Stanley Baldwin favoured the second scheme, but promised a 75 per cent contribution towards the cost of whichever scheme was agreed. Baldwin urged it upon the SR as a matter of national importance. The SR, after having a modification made which brought the proposed new terminus from Waterloo Junction (the former SECR Station) to the bank of the river, waited whilst the LCC prepared firm plans.

In 1929, the LCC sought powers to construct this on the basis that the SR would build a new terminus on the south bank on the site of the present Royal Festival Hall. Charing Cross Railway Bridge and Station would be demolished and replaced by a new road bridge. The government refused to make the 75 per cent grant towards the cost, which included reimbursing the SR for its outlays plus £325,000 compensation for disturbance.

Provisional agreement was reached between the SR and the LCC under which, for the former offering the latter the railway viaduct beyond York Road, the bridge and all the railway property at Charing Cross, the LCC offered the SR the eight and a half acre site of what is now the Royal Festival Hall but was then the Lion Brewery, which had been taken over by the brewers Hoare and Co, of Wapping. The new terminus would have been bounded by the Thames, Waterloo Road and the line of the railway viaduct. As well as the cost of the new terminus, the SR was to receive £325,000 as compensation for the loss of the Charing Cross Hotel and other losses. On 30 July 1929 the SR's directors recommended to the shareholders acceptance of the terms as the best available, but that the plan had been accepted with reluctance and then only because the government had said that it was of national importance. A substantial majority of the shareholders voted in favour of the agreement.

The LCC prepared a bill seeking the powers for a 95ft wide road bridge and approaches and railway access to the new station and new hotel site. The Old Vic Theatre would also have been demolished and rebuilt on a site almost adjoining. At this point it was the turn of parliament to influence events. Whilst the Commons Select Committee under Sir Henry Cautley could not find any fault with the idea of the road bridge, it could not stomach the idea of a railway terminus prominently placed at the bridgehead on the South Bank and so on 6 May 1930 the Bill was rejected.

Following this the LCC exhaustively examined a number of alternative proposals including constructing the new terminus below street level and out of site from across the river. In July 1931 the LCC still hoping to get its 75 per cent grant was ready to promote another Bill, but the Government had second thoughts and the Minister of Transport, Herbert Morrison, announced that in view of the controversial nature of the proposal for the railway terminus on the South Bank and the changes arising from railway electrification as well as the serious economic situation he was not prepared to renew the offer of a 75 per cent grant. This killed off the scheme.

## CHANGES TO THE SOMERSET AND DORSET JOINT RAILWAY

On 1 January 1930, the LMS took over the locomotives of the SDR whilst the rolling stock was divided between the two owning companies. On 1 July, the LMS assumed responsibility for the operating and commercial work of the railway and for the provision of traffic staff, whilst the SR took over all maintenance and civil engineering including signalling together with accounting work. At the same time, the distinctive blue livery of the SDR's passenger locomotives and carriages disappeared, but it had already disappeared from its goods locomotives.

Prior to the changes of 1930, the daily through train from Manchester to Bournemouth over the SDR had been given the name *The Pines Express* on 26 September 1927, named after the pine trees growing in the Chines in the Bournemouth area.

## A NEW COMPETITOR FOR CROSS CHANNEL TRAFFIC

Whilst the LNER was the SR's main competitor for traffic to the Continent serving Antwerp in Belgium and the Hook of Holland from Harwich, in 1927, a new competitor arose in the form of the LMS when in that year a service was started from Tilbury to Dunkirk operated on behalf of the Company by the French Angleterre-Lorraine-Alsace Societé Anonyme de Navigation (ALA). This was to fulfil an idea that the Midland Railway had had to develop a service from the Thames estuary to the Continent. The boat trains ran from St Pancras to Tilbury. The journey from London to Paris was not exactly the fastest taking about twelve hours compared to six and a half hours by the SR's route from London to Paris. The service suffered from unreliability especially through the winter fogs in the Thames estuary.

## THE CHANNEL TUNNEL

Although the Channel Tunnel was not built during the time of this book it does not mean that the threat of it to the SR's shipping did not exist.

Plans to build a cross-Channel fixed link appeared as early as 1802 when, a French mining engineer, Albert Mathieu-Favier, proposed a tunnel under the English Channel for horse-drawn coaches. Illumination would have been from oil lamps, and an artificial island positioned mid-Channel for changing horses.

In 1872 the Channel Tunnel Company was incorporated.

In the early 1880s work actually started on building a tunnel. Sir Edward Watkin, the Chairman of the Manchester, Sheffield and Lincolnshire, Metropolitan and South Eastern railways as well as being on the board of the East London Railway and the French Nord Railway, and Alexandre Lavelly, a French Suez Canal contractor, were involved in the Submarine Continental Railway Company and conducted exploratory works on both sides of the Channel. The work was carried out between June 1882 and March 1883. The tunnel was opposed by the British press and politicians and by the British military that feared that the tunnel might be used as an invasion route. In 1883 it was abandoned after just over a mile had been dug on both sides of the Channel.

In 1886, the Submarine Continental Railway Company acquired the 'rights and properties' of the Channel Tunnel Company and in 1887 assumed the name of that company.

At the Paris Peace Conference in 1919, the British prime minister David Lloyd George, as a way of reassuring the French about Britain's willingness to defend it against another attack by the Germans, repeatedly brought up the idea of a Channel Tunnel. Unfortunately, the French did not take the idea seriously and consequently nothing came of it, but that did not mean that the idea went away and it was looked into. Percy Tempest the Engineer and from 1920 the General Manager of the SECR, was the chief engineer for the work and in August 1923 he produced a memorandum on the subject. According to the memorandum, it was estimated that a Channel Tunnel would cost £29,000,000 to construct and based on the then traffic provide a return of over 5 per cent on that sum. The boring machine which would be used would provide a heading of 12ft in diameter at a rate of 120ft a day or a mile in 10 weeks and two such boring machines starting simultaneously from each side should meet in 2½ years which, allowing 6 months for the preliminary shafts and 18 months after the headings had met to complete the tunnel, it should be open in 4½ years. The subject was shelved. As to how the figure of the boring machine boring 120ft day boring a mile in 10 weeks was reached is uncertain as the actual figure is nearer a mile and a half in 10 weeks, so one must assume that some stoppage time for maintenance of the boring machine was allowed.

Winston Churchill was an advocate of the Channel Tunnel and published an essay in the *Weekly Dispatch* of 24 July on the subject entitled 'Should Strategists Veto The Tunnel?' In the article, Churchill vehemently argued against the idea that a Channel Tunnel could be used by a Continental enemy in an invasion of Britain

In 1929, a committee was set up by the British government to consider the construction of a Channel Tunnel, or rather fixed link to France. One idea which was rejected was for a bridge on the grounds that it would interfere with navigation, whilst other ideas were for submerged tubes. One idea which had been propounded by one William Collard in 1928 was for a tunnel and a new 253 mile long 7ft gauge electric line all the way from London to Paris with the journey taking 2¾ hours. The whole project was to cost nearly £200,000,000. Given the problems in the exchange of through passengers and goods by having a different gauge to that of the British and French railways, it was quite clear that this idea was impractical as it was less than 40 years since the final end of the GWR's 7ft gauge. The Committee rejected this, but reported in favour of the Channel Tunnel Company's scheme linking the SR to the French Nord Railway (NORD), the cost of which would be £25,000,000. It was estimated that the tunnel could be opened in 1938 and would electrically worked. The engineers addressed the concerns of both nations' military leaders by designing two sumps – one near the coast of each country – that could be flooded at will to block the tunnel.

Discussions between the SR and the NORD resulted in a firm proposal for which both railways were prepared to find the money. The British government of the day issued a White Paper which discussed the project in an inconclusive way. Matters were brought to a head on 30 June 1930 when a Resolution in the House of Commons was moved by Mr E. Thurtle, a Labour member and seconded by Sir Basil Peto, a Conservative. In the ensuing discussion, the speeches on both sides of the Commons were virtually confined to exposing prejudices and preconceived ideas and there was no real discussion on the economic issues or the financial prospects. Opponents of the project included the Conservative Sir Samuel Hoare, who previously had been Secretary of State for Air, and Herbert Morrison, the Labour Party Transport Spokesman, whilst support came largely from Labour Party back-benchers. The Prime Minister, James Ramsay MacDonald, made a lengthy speech, which was rather woolly and which indicated that he felt too many doubts and saw too many difficulties to support the Channel Tunnel. On a free vote there were 172 for the resolution and 179 against; the resolution was defeated which put paid to the Channel Tunnel for the foreseeable future as parliamentary approval was needed for legal and legislative formalities involved including a treaty with France.

Had the Channel Tunnel been built, it is clear that through services would have been operated jointly by the SR and the NORD and given that the NORD never electrified any of its services, one wonders what electrical system would have been used through the Channel Tunnel.

## THE CONTINUING THREAT FROM AIR TRAVEL

To counter the rising threat from air lines on the London to Paris route on 13 September 1926, the NORD had introduced an all-first-class Pullman service between Paris and Calais named the *Flèche d'Or*. On 1 May 1927, Imperial Airways introduced its *Silver Wing* service on the London–Paris route in direct competition. The service was operated with a dedicated fleet of three Armstrong Whitworth Argosy's, named the *City of Birmingham, City of Glasgow* and *City of Wellington*. The aircraft were painted silver externally and also had silver and grey cabin interiors. New more comfortable seating than that previously provided with shoulder and head rests was also installed. The flight left London at noon each day. On the two hour and 30 minute flight, a steward would serve a four course lunch and offer a bar service to the twenty passengers on board the aircraft. The SR's response took slightly longer than might have been expected. On 15 May 1929, the SR introduced the all first class *Golden Arrow* Pullman train between London Victoria and Dover, while simultaneously launching a new first class only ship, the Canterbury, for the ferry crossing. The journey from London to Paris including the sea crossing took 6 hours and 35 minutes. Whist the rail and sea route was slower than the air route, what was not mentioned was that the airport for London was at Croydon and that for Paris at Le Bourget and neither were in the centre of the respective capital cities, whereas Victoria and the Gare du Nord were.

According to the *Daily Chronicle* of Saturday 18 May 1929:

NEW LUXURY LONDON- PARIS SERVICE CROSS-CHANNEL BOAT WITH GLASS-SHELTERED DECK. The luxury of modern travel is typified in the new "*Golden Arrow*" all Pullman service between London and Paris. You can recline in an armchair for the whole of the journey, if you so desire, for the new cross-Channel steamer, *Canterbury*, which connects the English and French trains at Dover and Calais, has a glass-sheltered awning deck on which armchairs take the place of the traditional canvas and basket seats. Owing to the fact that no more than 300 people are carried the trains get away so quickly after the arrival of the steamer that the journey between the' two capitals takes only six hours 35 minutes. Compared with the old *Golden Arrow* schedule that is a saving of 20 minutes from London to Paris and 40 minutes from Paris to London. The *Canterbury* is the largest and most luxurious steamer on the cross- Channel service and is oil-driven, her tonnage is 3,100.

In referring to the 'old Golden Arrow', the newspaper is referring to the French train as an article in its edition of 13 September 1926 it refers to the train as such.

This though is not quite the end of the story. In 1929, the Big Four railway companies obtained general powers to operate air services. The result of this was that Walker entered into negotiations with Imperial Airways which ranged from investment to mutual help and that same year, the Southern acquired shares in Imperial Airways. On 12 February 1930, the SR's board authorised Walker to sign the memorandum agreeing terms with Imperial Airways.

## THE CONTINUING THREAT FROM BUSES AND COACHES

To counter the continuing threat from road competition, the SR invested in some of the bus companies. According to Bonavia, by the end of 1930 the SR had acquired substantial holdings in: Hants and Dorset Motor Services, Thames Valley Traction Company, Devon General Omnibus and Touring Company, Wilts and Dorset Motor Services, Southern National Omnibus Company, Aldershot and District Traction Company, East Kent Road Car Company, Maidstone and District Motor Services, Southdown Motor Services and Southern Vectis Motor Services. The Southern's participation in Devon General and Thames Valley was joint with the GWR. Southern Vectis, on the Isle of Wight, was a special case. It had been founded 1921 in Cowes as Dodson and Campbell, and in 1923 the company was renamed the Vectis Bus Company. In 1929, the company was purchased by the SR and was incorporated as The Southern Vectis Omnibus Company Limited. The Southern then transferred half the equity to the Tilling Group, but the name Southern Vectis was retained. Southern National came out of a split of the National Omnibus and Travel Company. There was also a Western National which the GWR invested in and an Eastern National in which the LNER and LMS invested. Apart from these, there was the London General Omnibus Company which was part of the Underground Group and operated bus services in London, and the East Surrey Traction Company which operated services in Surrey, Kent and Sussex and also Autocar which operated in the same area as East Surrey.

By the summer of 1930, the SR faced competition from daily express coach services operating from London to Seaford, Rye, Maidstone, Gosport, Crowborough, Winchester, Tenterden, Aldershot and Farnham, the Medway Towns and Sheerness, Weymouth, Dover, Folkestone, Littlehampton and Bognor Regis, Eastbourne, Worthing, Hastings, Brighton, Devon and Cornwall, Southampton and Bournemouth, Portsmouth and Southsea and the Isle of Thanet (Herne Bay, Margate, Ramsgate etc). By having investment in the bus companies which also operated the coach services the Southern was able to have some influence in them.

In 1930, John Elliot was made assistant traffic manager for the SR and Walker put him on all the road transport boards that the SR had an interest in. In doing so, Walker told him that when he went to a road transport board meeting, he represented the SR's investment in that company and not to go as a railwayman, which instruction Elliot followed with good results.

Whilst coach travel was cheaper, trains could offer carriages with lavatories and also provide refreshments without the need to provide stops for the aforementioned and they were also faster. In 1924, new up to date corridor trains were built for the Kent Coast services, whilst 1925 had seen a standard design of coach started to be produced in quantity for the West of England and Brighton sections. The Charing Cross and Cannon Street to Hastings services were given new trains in 1929, which had to be built to a width of only 8ft because of loading gauge restrictions in the five intermediate tunnels between Tonbridge and St Leonards.

8 July 1929 saw a named train inaugurated on the Waterloo to Bournemouth service. This was the *Bournemouth Limited* which ran non-stop between Waterloo and Bournemouth in two hours. The inaugural train was hauled by Lord Nelson class 4-6-0 No E850 *Lord Hawke*. A trial run for the railway's directors, officials and the press was held on Thursday, 4 July. On the trial run, *Lord Hawke* had reached 85mph. The train left Bournemouth at 8.40 in the morning and Waterloo at 4.30 pm in the afternoon and included a portion from and to Weymouth.

## CAR FERRY

The SR was quite happy to make money from motorists and in July 1927 started a car ferry to the Isle of Wight. The *West Sussex Gazette* of 21 July 1927 reported:

CAR FERRY. The first motorcar ferry of its kind established in Great Britain is starting between the Isle of Wight and old Portsmouth. It is being run by the Southern Railway Co., who have had built on the Clyde a specially-designed, motor-driven ferry vessel, capable of carrying twenty cars and one hundred passengers. This boat will ply between Portsmouth and Wootton, Isle of Wight and carry only motor-cars and their occupants. By means of it motorists will be able to get to and from the island with their cars quite easily, the passage taking as hour each way.

Townsend Brothers tried a car ferry between Dover and Calais starting on 7 July 1928, but this was suspended after 31 December 1929 but recommenced on 15 April 1930.

## THE EAST KENT LIGHT RAILWAY

On the East Kent Light Railway the passenger service between Eastry and Sandwich Road was withdrawn on 1 November 1928. For some years the line was still shown in Bradshaw but without a train service. The stations concerned were Poison Cross, Roman Road and Sandwich Road.

## THE SOUTHERN HEIGHTS LIGHT RAILWAY

The SR nearly gained a new railway to work. This was the Southern Heights Light Railway. The original proposal went back to 1898 when there was a proposal to build a railway from Orpington via Cudham to Tatsfield. The engineer of the railway was Colonel Holman Frederick Stephens who was involved in the engineering and the building, and later the managing of sixteen light railways in England and Wales, including the Kent and East Sussex and the East Kent light railways. The Orpington, Cudham and Tatsfield Light Railway obtained its Light Railway Order, but unfortunately investment in the railway did not materialise and so it was never built.

The plan was revived but on a more ambitious scale in the mid-1920s. What gave the scheme the impetus was the extension of the SR's electric trains to Orpington on 12 July 1925. *The London Gazette* of 27 November 1925 recorded an application for a Light Railway Order for a 15¼ mile long light railway from a junction 180 yards south of Orpington station to a junction half a mile south of Sanderstead station. The promoters were Charles Igglesden, who was an author and journalist, Jeremiah MacVeagh, Holman Fred Stephens and others. The application was signed by Stephens.

The railway was to have been a single track railway with passing loops. The intermediate stations would have been at Mitchley Wood,

Hamsey Green, Chelsham for Warlingham, Tatsfield, Westerham Hill, Cudham and Biggin Hill, Downe and Keston and Green Street Green for Farnborough. Unlike other lights railways which Stephens was involved with, there were to have been no level crossings at all on the line, which would have required twenty-three bridges to have been built. Construction of the line would have required the excavation of 631,000 cubic yards of material.

*The Westminster Gazette* of Monday 7 December 1925 carried an article which read:

NEW KENT RAILWAY. Plans of the 'Southern Heights' Light Railway scheme, to cost £511,000 between Orpington and Sanderstead, have now been lodged with the local councils concerned, and will be considered by Farnborough Town Council to-night. It is intended to have electric passenger and steam goods trains, with three stations and several halts.

At the meeting Farnborough Council opposed the scheme. A public inquiry into the Light Railway was opened at Orpington Village Hall on 3 March 1926. This brought the Ministry of Transport's conditional approval for the Light Railway.

Prior to the inquiry on 24 February, Walker had recommended to the SR's directors that the Company work the Light Railway, to guarantee interest on its debentures not exceeding 5 per cent on £300,000, to obtain powers to take up capital and guarantee debentures and to electrify the railway to Southern standards. On 3 March, Walker recommended that the electrification capital of £140,000 be provided by the SR.

From the autumn of 1926, the Light Railway made its appearance on SR official maps including those in carriage panels above the backs of the seats.

In 1927, the Light Railway's debentures under guarantee by the SR were increased to £330,000 and in 1928 the scheme crystallised. The

Southern Heights Light Railway was to acquire the necessary land and build it at its own expense whilst the SR would work it at 75 per cent of the gross receipts, would electrify it at a cost of £140,000 and would be repaid in cash or in shares with interest at 5 per cent. The SR's guarantee of interest was confirmed to be 5 per cent on £330,000 of debenture stock and whilst the liability was at £10,000 or more the SR would appoint two directors to the Light Railway's board. The ordinary capital was intended to be £500,000.

In April 1928, Walker led an inspection party over the route of the Light Railway which comprised senior officers of the SR and Stephens. According to Alan A. Jackson in *London's Local Railways* (David and Charles 1978) it was decided that at Orpington, the connection between the SR and the Light Railway would be with the Up and Down local lines at the country end of the station and not with the Up fast line as originally intended. In order to assist the working of the Light Railway, it would be necessary to provide facilities at Orpington for the splitting and the joining of the multiple unit electric trains. The men from the Southern were content with the station sites that had been selected by Stephens. There would be passing loops at Hamsey Green, Tatsfield and Westerham Hill and the stations would consist of island platforms and buildings of a permanent character, whilst at the other stations the buildings would have been of a portable nature to minimise expenses should the railway be doubled. The station buildings would include shops to generate extra revenue where conditions were favourable. According to Jackson, one gets the impression that the SR was out to change the nature of the still largely remote and beautiful area for the benefit of its shareholders, but with the minimum outlay of its own capital.

On 29 December, the Minister of Transport, Herbert Morrison, approved the Southern Heights Light Railway Order. The Light Railway Order was for a slightly revised scheme for a 16 mile line over the same route. Its capital was to be £593,890 of which construction costs were estimated as requiring £521,574: the extra expense being to meet the requirements of the Southern.

The February 1929 edition of the *Railway Magazine* recorded – 'The Minister of Transport has recently made the following Order: – The Southern Heights Light Railway Order 1928, authorising the construction of a light railway in the counties of Kent and Surrey running from Orpington to Sanderstead.' According to its May edition, the SR had agreed to work the Light Railway for 75 per cent of gross receipts guaranteeing interest at a maximum on the debenture stock up to £330,000. The magazine said that the electrification would cost £140,000 which the Southern Heights Light Railway would meet by payments either in cash or in fully paid-up preference shares.

Things started taking a different turn from what Stephens wanted. At a meeting in May 1929 that was attended by Walker and the SR's chief officers, Stephens agreed to the plans for the connection at Orpington where the SR planned a siding parallel with the Down main line for the stapling of electric stock and the SR's Assistant Engineer (New Works) was instructed to start dumping material as and when it was convenient to form the necessary widening of the embankment for the junction with the Light Railway, whilst it was also agreed at the meeting that the Light Railway's bridge under the main line should be constructed to allow double track later. Stephens successfully resisted attempts to persuade him to find another £50,000 to have the whole railway built to allow double track later, but he did however agree to consider whether provision should be made for double track between Orpington and Green Slade Green and perhaps Sanderstead and Chelsham. The SR's men also insisted on improvements to certain curves, on fencing to the standards of the company's electric lines and on the normal SR main line type ballasting. Stephens, who was used to lightly laid unfenced lines with stations built of corrugated iron and second, third and fourth hand steam locomotives and petrol engine railcars, objected to this extravagance. It is clear that the Light Railway was ceasing to be a light railway and Stephens' protests were quite understandable as he was having great difficulty in raising the unsecured capital. In 1930, he made an unsuccessful fund raising trip to America. As the capital could not be raised, the Light Railway Order lapsed in December 1930 and in January 1931, as a result of this, a new Light Railway Order was applied for and authorisation was sought to deviate from the authorised route of the Light Railway in an effort to reduce construction costs by £17,245. The line was to have taken a different route in the parishes of Cudham, Tatsfield and Titsey, on the Kent/Surrey border. On 23 October 1931 Stephens died at 62 and the management of his railways was taken over by his former 'outdoor assistant', William Henry Austen. With Stephens' death the scheme withered and faded away, but even before that it had effectively been killed when on 23 July that year it was reported to the Southern's Board that the Light Railway's powers had lapsed and in view of the forthcoming London Transport authority (London Passenger Transport Board), the Board decided not to support the application for a new Light Railway Order and the scheme was deleted from the company's publicity. Despite this, Stephens told Klapper shortly before he died that he was raising the last £400,000 in America and that he would soon be making a trip to clinch the deal. The application for the further Light Railway stood on the books until 1932, when the Ministry of Transport ruled that there would be difficulty in making any Order which 'might be inconsistent with the London Transport co-ordination scheme in the London Passenger Transport Bill'.

The only construction for the Light Railway that was ever having started was south of Orpington station on the east side of the line where during the 1930s some excess spoil was delivered to the site ready for construction.

## THE ROMNEY, HYTHE AND DYMCHURCH RAILWAY

The idea for a railway from Hythe to New Romney went back to 1884 when the SER obtained powers to build a railway from New

Romney to Hythe. Nothing came of this. Nor did a Hythe and New Romney Light Railway which obtained its order in October 1900.

The idea for the Romney, Hythe and Dymchurch Railway (RHDR) originated with two wealthy motor racing drivers – Captain John Howey and Count Louis Zborowski – who were both interested in 15in gauge railways and had a dream of building a main line railway in miniature to that gauge. There was already in existence a public carrying 15in gauge railway – the Ravenglass and Eskdale Railway in Cumberland, which had taken over the defunct 3ft gauge railway of that name and reopened it on the narrower gauge in 1915. Both Howey and Zborowski thought about buying the Ravenglass and Eskdale Railway and extending it. Nothing came of this. Zborowski at his home, Higham Park at Bridge in Kent, had constructed a 15in gauge railway and agreed to donate the rolling stock and infrastructure to the projected main line in miniature. Prior to the First World War, Howey had had a 9½in gauge miniature railway at his home at Melford Grange, near Woodbridge in Suffolk. He later moved to Staughton Manor in Huntingdonshire. There he built another miniature railway, which was later relaid to 15in. Howey had joined the Royal Flying Corps at the start of the war, but during it he was captured by the Germans and taken prisoner in 1915. After the war he had abandoned his railway interest for a time. It was his meeting with Zborowski that rekindled it.

With Zborowski having already a San Pareil 15in gauge 4-4-2 or Atlantic type locomotive, he wanted something that would be the biggest and best miniature locomotive ever built. He went to Henry Greenly who was the best miniature locomotive designer of the early twentieth century. Not one, but two locomotives were built by Davey and Paxman of Colchester – one for each of them. At Zborowski or Howey's insistence the locomotives were built to resemble Nigel Gresley's A1 4-6-2s or Pacifics on the LNER which were amongst the most impressive, if not necessarily the best, British express locomotives of the day.

Unfortunately Zborowski was killed in a motor racing accident at the Monza Grand Prix in Italy on 19 October 1924.

John Snell in *One Man's Railway* (David and Charles 1993) says that Zborowski's death would have stopped Howey's return to miniature railway world. Two things happened though. His father had died a few months previously and he was now his own master as far as finding large sums of money was concerned. The other was the encouragement of friends who shared his interest. Howey's mother was all in favour of encouraging his interest in railways as they were a lot safer than racing cars. There was also the slight matter that Davey and Paxman were pressing on with building the two locomotives and that they would soon be ready. Also, Howey felt that the main motive for pressing on was that the railway would be a memorial to Zborowski.

In early 1925, Greenly was given the task of finding a suitable location to build it. The railway would have to be the best miniature railway in the world. This meant that it had to be at least 7 miles long and somewhere straight and level so that the trains could go really fast. Greenly looked at a number of locations. One was at Brean Sands in Somerset between Burnham-on-Sea and Weston-super-Mare. The idea was to run from Burnham-on-Sea via the GWR Station at Brent Knoll to Weston-super-Mare. The line would have been just over 7 miles. Whilst the area was becoming a place for summer holidays, there was not much public benefit. Nor was the GWR happy, as it did not want competition. Greenly corresponded with Felix Pole, the GWR's General Manager, who did not give him any encouragement. Next the West Sussex Light Railway from Chichester to Selsey was looked at.

This railway was 7¾ miles long and was ideal, and whilst the railway may have been struggling it was still operating and local people would certainly not have been happy exchanging a full size railway, however decrepit, for a new miniature railway of which they knew nothing. The railway had a large number of road crossings which would have been prohibitively expensive to either gate or bridge and there was the slight matter of Colonel Stephens, who was the manager, receiver and engineer and who was not easy to convince, so nothing came of that idea.

Sir Herbert Walker was encouraging and suggested Romney Marsh, where particularly in Dymchurch there was still feeling in favour of a railway. The indications were that, whilst a full size one would not pay, a smaller gauge one would. Walker thought that that might just be the place for the main line in miniature and Greenly went to New Romney and had a look. The route proposed was 8 miles long, level and without any important obstacles. He reported favourably to Howey on Walker's proposal, who visited New Romney on 8 September 1925 and agreed there and then that that was the ideal location for his proposed railway.

Even before Howey had visited New Romney, wind of what was proposed got out. The *Kentish Express* of Saturday 29 August 1925 reported that at the meeting of Romney Marsh Rural District Council on 27 August:

> A letter was read from Mr. Henry Greenly, of Farnborough, stating that he had a client who was considering the building of a light railway from New Romney to Hythe, and asking if the Council had objection to the scheme, stating that he had the moral support of the Southern Railway Company. The Council decided to raise no objection to the scheme.

In order to build the light railway a light railway order was required and this was applied for on 14 November 1925. The hearing for the Light Railway Order was held on 15 and 16 January 1926 at the Assembly Rooms in New Romney before the Light Railway Commissioner, Alan D Erskine, who had previously composed the draft order. Support for the railway came from the local councils and opposition from land owners and the local bus company – the East Kent Road Car Company.

On 19 February 1926, the Minister of Transport Wilfred Ashleym indicated his intention to approve the application for the order. On 26 May, the Romney, Hythe & Dymchurch Light Railway Order 1926 was made which incorporated the Romney, Hythe & Dymchurch Light Railway Company as a statutory public utility undertaking with powers to construct and work the proposed railway and also included compulsory purchase powers over the land required: these powers had to be used to acquire six plots of land on the proposed route. Construction of the railway started soon after and in fact even before the Light Railway Order hearing, Howey had started building the railway on some land that he owned at New Romney in December 1925. Construction of the railway was placed in the hands of Greenly.

The original idea had been for a line with passing loops, but during the summer of 1926, Howey told Greenly that he wanted double track from New Romney to Dymchurch and when construction had reached Dymchurch he wanted it to Hythe.

The original intention had been to open the railway from New Romney to Dymchurch in the summer of 1926, but because of a delay in starting work on a bridge at The Warren this did not materialise. There was however a royal train as on 5 August when the Duke of York, later King George VI visited a boys' camp at Jesson, of which he was the patron. The *Daily Chronicle* of 6 August carried an article headed 'DUKE OF YORK AS ENGINE DRIVER' and which began, 'During his visit to the Boy Scouts' camp at New Romney yesterday the Duke of York drove one of the miniature engines, with many of the guests as passengers, on the light railway which is being constructed from New Romney to Hythe.'

The railway finally opened to the public on 16 July 1927 with the opening ceremony being performed by the Lord Warden of the Cinque Ports, William Lygon, 7th Earl Beauchamp. The Daily Chronicle of 16 July carried an article which began "TOY" RAILWAY OPENS TO-DAY. Seven Stations on Nine Miles Track.

SMALLEST IN WORLD "The smallest public railway in the world," the Romney, Hythe and Dymchurch Railway, will be opened by Earl Beauchamp, Lord Warden of the Cinque Ports, this afternoon. The ceremony will take place at the West Hythe station of the railway. From to-morrow a regular service will be run every 40 minutes, the stations being :— Hythe, Prince of Wales, Botolphs Bridge, Dymnchurch, Holiday Camp. Romney Warren, and New Romney. Altogether there are nine miles of track. Though essentially a miniature railway—the stations are so small that one stumbles across them rather than notices them—it is quite up to date.

*The Daily News* of 18 July carried an article which read:

KENT'S TOY RAILWAY. LITTLE LINE ALONG THE COAST OPENED. One of the smallest public railways in the world was opened at Hythe on Saturday. when Earl Beauchamp, Lord Warden of the Cinque Ports, pressed a button to regulate a signal permitting the departure of the first train on the Romney, Hythe and Dymchurch Light Railway, the new line with a gauge of 15 inches. The Mayors of the Cinque Ports, with their councils and officials, travelled on the train, which conveyed 100 people in coaches to Romney in less than 30 minutes. At every road and crossing crowds of people thronged to cheer the brave little train, and public and private cars raced to give their passengers the opportunity of seeing the passage of the eight-ton engine Samson and its Lilliputian carriages. General Sir Ivor Maxse, the chairman of the railway company, described it as a sporting railway, because it was built by sportsmen. The distance between Hythe and Romney is nine miles. The new railway is not quite the smallest public railway. The Eskdale Railway, which runs from Ravenglass (in Cumberland) to Boot, at the foot of Scawfell (Scafell) is also of 15 inch gauge. It is only 7¼ miles long. Its timetable duly appears in "Bradshaw."

Even before the railway was completed Howey had plans for an extension of 5½ miles from New Romney to Dungeness. The line was to be double track throughout, but with a balloon loop for turning the trains at Dungeness. Another Light Railway Order was applied for and a public inquiry was held at New Romney on 18 April 1928. On 12 July 1928 the Romney, Hythe & Dymchurch Light Railway (Extension) Order was granted, but even before the Order had been granted the line between New Romney and The Pilot had opened on 24 May 1928. Turning was done by means of a temporary triangle. The rest of the line through to Dungeness opened on 3 August 1928.

Motive power initially consisted of five 4-6-2 locomotives modelled on the LNER's A1 Class locomotives, two 4-8-2 locomotives of British outline and one 0-4-0 tender tank locomotive and which were joined in 1931 by two 4-6-2s of Canadian outline. Coaching stock initially consisted of four wheeled carriages, but soon bogie carriages were also provided and eventually the four wheeled carriages were scrapped. There were also goods vehicles and some freight was carried.

## ACCIDENTS

In 1927 the SR suffered three serious accidents, all involving 2-6-4 side tank locomotives of Classes K and K1. Class K originated with a prototype built by the SECR in 1917 and was multiplied after the grouping and Class K1 was a single three cylinder version of Class K which was built in 1925.

The Ks were put to use on trains on the former SECR and LBSCR lines including expresses, whilst the K1 worked only on the former SECR lines, but because they were tank engines they were not able to work on the longer routes of the former LSWR

On well-maintained track, the K class proved successful, but the entire track on the former SECR and LBSCR sections was not all well

**Maunsell's River** Class 2-6-4Ts had the misfortune to roll heavily on track that was not well maintained and gained the nickname 'Rolling Rivers'. Following the Sevenoaks disaster involving a member of the class they were rebuilt as 2-6-0s. Here is No A807 *River Axe*. *John Scott-Morgan collection*

laid and the former particularly so. Engine crews complained that the engines rolled heavily and unpredictably and they gained the nickname 'Rolling Rivers'. This was also in part due to a design fault.

The first accident occurred on 31 March 1927 near Wrotham on the former LCDR line from London to Ashford via Swanley and Maidstone when Class K1 No A890 *River Frome* derailed itself at a level crossing when the flanges of the lead coupled wheel mounted the rails whilst travelling at about 60 mph, fortunately without serious injury to the crew and passengers and the engine and carriages were only slightly damaged. The accident was blamed on the track and until that could be rectified the K and K1 classes were banned from the line. This was done quite quickly as on 20 August, the same engine on the same line was derailed at Bearsted when the lead driving wheel mounted and completely dropped off the rails at 40mph, derailing the train and causing serious damage to the track. In the accident there were no fatalities, but six people were injured. The cause of the accident was put down to a defect in the track and poor rolling stock stability.

Four days after the Bearsted derailment there occurred a much worse derailment at Sevenoaks on the former SER main line from to Dover via Tonbridge, Ashford and Folkestone. Class K locomotive No A800 *River Cray* whilst hauling the 5.00 p.m. train from Cannon Street to Deal which contained the Pullman Car *Carmen* was travelling at about 60mph between Dunton Green and Sevenoaks Stations, when the whole train became derailed at a spot about three-quarters of a mile on the London side of Sevenoaks at about 5.30 pm, the spot in question being in a cutting and spanned by a bridge. The cab of the engine struck the bridge and the engine was turned on its side across the cutting. The leading coaches piled up against it, killing 13 and injuring 132. The driver and fireman survived.

*The Daily Chronicle* of 25 August reported on the accident in an article which had the headlines: THIRTEEN DEAD IN KENTISH RAILWAY DISASTER LONDON EXPRESS WRECKED DEAL TRAIN DERAILED OUTSIDE SEVENOAKS

*The Daily News* of the same day reported on the accident in an article which had the headlines: 11 KILLED AND 60 INJURED IN EXPRESS DISASTER BUSINESS TRAIN CRASH VICTIMS MOSTLY IN PULLMAN CAR "LIKE BATTLEFIELD"

The *Locomotive Magazine* of September 1927 carried a short article which read:

Sevenoaks Accident, Southern Ry .—On Wednesday evening, August 24, the 5 p.m. train from Cannon Street to Deal was derailed at 5-30 p.m. at Riverhead, between Dunton Green and Sevenoaks. The train was headed by the 2-6-4 tank engine

*River Cray*, No. A800, and consisted of seven bogie carriages and a Pullman car. Thirteen passengers lost their lives, and 48 other passengers were more or less seriously injured. Sir John Pringle (the, Chief Inspecting Officer of His Majesty's Railway Inspectorate) is conducting an enquiry into the cause of the disaster on behalf of the Ministry of Transport.

At the inquiry, other drivers testified about the instability of the class and it emerged that one locomotive had previously derailed at speed, but in that instance, it had re-railed itself. Pringle concluded that the engines' high centre of gravity, their hard springing, and the tendency of the water in the side tanks to surge, all caused the engines to roll dangerously at speed, so much so that in this accident the nearside wheels had lifted.

The report of the accident was not published until 1928. *The Daily News* of 2 February carried an article which began:

SEVENOAKS DISASTER. BAD TRACK THE CAUSE. INQUIRY REPORT. HOW ROLLING ENGINE WAS DERAILED. Serious reflections on the condition, at the time, of the Southern Railway track are made by Colonel Sir John Pringle in his report on the Ministry of Transport inquiry into the disaster near Sevenoaks. Kent, last August. [and went on to quote him] 'In my opinion the cause of the derailment must be attributed to the rolling movement of engine No 800, initiated, possibly in the first instance when it passed over the trailing connection on the down line at 60 m.p.h. I find the condition of the road, in respect of foundation and maintenance, was the initial cause of the rolling motion.'

According to Elliot, Sir Herbert Walker, who had been on holiday in Switzerland on his return called a meeting of the Chief Officers concerned. He said that there had been a serious accident and that something was wrong somewhere. He asked Maunsell if he thought that his tank locomotive had become unstable. Maunsell replied that it was perfectly stable and blamed it on the track. Ellson, the Chief Civil Engineer, said that there was nothing wrong with the track. Walker said that he suspected that the locomotive became unstable on the track and ordered Maunsell to withdraw all the 'Rivers' from service that day, with the possibility of redesigning them as tender engines and Ellson to let him have an estimate of the cost of re-ballasting and perhaps re-laying the SER main line from London to Dover.

Following the accident, trials were carried out on the LNER's former Great Northern main line under the supervision of that company's Chef Mechanical Engineer, Nigel Gresley, to gain an unbiased review of their riding qualities. The locomotives involved were K Class No A803, K1 Class No A890 and King Arthur Class No E782. No problems were reported. Walker then ordered further trials to be held led by the engineer Sir John Aspinall on the SR's Western section main line between Woking and Walton. The trials were terminated by the SR's Operating Department, as the riding of the locomotives at speeds near 80mph rendered the locomotives unsafe.

Before the results of the trials had been published, all River class engines had been rebuilt, becoming the first twenty of the U Class 2-6-0 tender engine design and the prototype U1 Class 2-6-0 tender engine design.

The Sevenoaks accident had repercussions beyond the SR. On the LNER's former GER route to Southend, passenger services were in the hands of a variety of locomotives – principally express tender locomotives such as the Claud Hamilton Class 4-4-0s and the 1500 Class 4-6-0s. The LMS' former London, Tilbury and Southend Railway section used tank locomotives on its services to Southend. In 1927, when additional motive power was required by the LNER for its service to Southend, a 2-6-4 side tank locomotive of a power approximately equal to that of a 1500 Class was designed .According to F.A.S. Brown in *Nigel Gresley – Locomotive Engineer* (Ian Allan 1961), unfortunately the derailment of a River Class 2-6-4 side tank locomotive on the Southern Railway at Sevenoaks which resulted in the deaths of thirteen passengers threw doubts on the wisdom of using such engines on fast passenger trains. Although the press criticism was unjustified as the track bed of the former GER was somewhat better than that of the former SER, the design was abandoned and in its place ten new 1500 Class locomotives were built. On the LMS a new class of 2-6-4 side tanks was put into service in 1927 following a reassurance from the Ministry of Transport that they were perfectly alright provided that they were not used on fast services!

## MOTIVE POWER

Five new classes of steam locomotive were put into service up to the end of 1930. In 1921, Urie on the LSWR had put into service on that railway four Class H16 4-6-2 side tanks for interchange goods traffic both within that railway and with other railways. He also put into service four Class G16 4-8-0 side tanks for shunting at Feltham yard and short distance freight duties in the London area. More of these were on order in 1922, but Maunsell postponed the order and replaced it with one for eight three-cylinder 0-8-0 side tanks of Class Z. These were all built at Brighton in 1929 and proved to be very useful locomotives. The *Locomotive Magazine* of April 1929 reported 'These engines are intended for service in the principal goods yards and sorting sidings of the Southern Ry., and, whilst powerful enough to operate trains over the humps, have sufficient play for the leading and trailing wheels to permit of the engines traversing curves of 4½ chains radius.'

1930 saw the arrival of Class V three cylinder 4-4-0s, known as the Schools Class. Their intended role was intermediate express work, but they proved capable of not only that work, but also top express work. They had a round top firebox which was not divided like the Lord Nelson's. The boiler was a shortened version of that used on the King Arthurs. They proved popular with engine crews and could work most routes other than those west of Exeter but including the Tonbridge to Hastings line with its restricted loading gauge necessitating special rolling stock. *The Locomotive Magazine* of April 1930 carried an article which began:

> Southern Ry, New 4-4-0 Express Locomotives TEN three-cylinder 4-4-0 type engines are being built by the Southern Ry. at their Eastleigh works, to the design of Mr. R. E. L. Maunsell, Chief Mechanical Engineer. Officially known as the "V" class, the engines will bear names of the best known public schools in the Southern Counties. The first, No. E900 is named *Eton*.

A total of forty of the locomotives were built between March 1930 and July 1935.

In 1927-9, of ten of William Stroudley's LBSCR Class E1 0–6-0 side tanks, which were first built in 1874, were rebuilt as 0-6-2 side tanks.

**The Schools** Class 4-4-0s were Maunsell's masterpiece and were able to work over a large portion of the Southern including the Tonbridge to Hastings line with its restricted loading gauge. Here is No 900 *Eton* alongside King Arthur Class No 800 *Sir Meleaus de Lile*. John Scott-Morgan collection

The purpose of the rebuild was to provide additional motive power for services in the West Country without the need to build new locomotives similar to the 0-6-2 side tanks of the former Plymouth, Devonport and South Western Junction Railway. By using some surplus Class E1 locomotives the bother of scrapping them was also saved. The Class was reasonably successful. The rebuilt locomotives were classed E1R rather than E7 as was originally intended.

Classes U and U1 arose from the decision following the Sevenoaks accident of 21 August 1927 to rebuild the K Class 2-6-4 two-cylinder side tanks and the K1 Class 2-6-4 three-cylinder side tank as 2-6-0 tender locomotives. All twenty Class Ks were rebuilt to Class U in 1928, whilst a further twenty which were under construction were altered to Us prior to their completion in 1928-9. A further ten were built new in 1931. The solitary Class K1 was rebuilt as Class U1 in 1928, and a further twenty U1s were built new in 1931. Both Classes U and U1 proved very capable even to working express trains.

*The Locomotive Magazine* for April 1929 carried an illustrated article on Class U which began:

NEW PASSENGER LOCOMOTIVES—SOUTHERN RY. THE 2-6-0 type of locomotive has hitherto been used as a goods or mixed traffic engine, being a development of the 0-6-0 type. Its adoption as a purely passenger locomotive by the Southern Ry. is a new departure, and has been brought about by the provision of larger coupled wheels, 6 ft. 0 in. in diameter, an increase of 6 in. This enables the engines to be put to a large range of duties in passenger working, as they are suitable for a maximum speed of 70 m.p.h., while their large adhesive weight enables them to take longer and heavier trains than the 4-4-0 type. Consequently, they can deal with all but the fastest and heaviest services which require a heavier engine of the 4-6-0 type with large wheels. Ten of these 2-6-0 type passenger engines have been constructed at the Southern Ry. Co.'s works at Brighton, while another ten

**The Class U** 2-6-0s were a mixture of rebuilds from the River Class 2-6-4Ts and new builds and proved to be very useful locomotives. Here is No 1798 which was originally built in June 1925 as No A798 *River Wey* and rebuilt to its later form in August 1928. *John Scott-Morgan collection*

are under construction at the company's Ashford works. They are known as the "U" class. The design closely follows that of Mr. Maunsell's "N" class, 2-6-0 type, mixed traffic engines, nearly the whole of the parts—boiler, cylinders, motion, frame details, valve gear, etc., being identical, the only difference being in the wheels and framing.

On 29 December 1929, Class U No A629 left Ashford works fitted with German designed equipment for burning pulverized fuel which was stored in a bunker on the tender. Experience with the type of fuel in Germany and New South Wales had been quite encouraging, but they had used soft brown coal which was easier to crush into powdered form than the harder coal used in Britain, which, whilst crushable, was liable to produce lumps which jammed the mechanism feeding it into the firebox. Following test runs around Ashford until the end of January 1930, the locomotive was transferred to Eastbourne for trials with classmate No A627 for which both locomotives were fitted with speedometers. At Eastbourne, a special bunker was provided to contain the pulverized fuel, but the trials were not a success. The locomotive's progress through the Sussex countryside was marked by dense clouds of dirty smoke, lineside fires and the sound of fire engines bells. The special bunker at Eastbourne once displayed distinct signs of being about to explode and the fuel had to be hurriedly emptied into a number of wagons and when as much fuel as possible had been removed the bunker lid was opened and the blowers turned on to remove any remaining dust pockets, which promptly burst into flame and with a thunderous crash a cloud of dense black smoke shot into orbit over Eastbourne to the dismay of hoteliers, holiday makers, housewives and railway officials. The equipment was removed at the end of 1932. The locomotive only ran 3,497 miles burning pulverized fuel and any savings gained were expended by the crushing and the need to compensate farmers for lost crops. Prior to the grouping, the Great Central Railway had experimented with the use of pulverized fuel but had abandoned it on the cost of preparing the fuel and the provision of grinding mills, plus the difficulties of storage.

An interesting locomotive was built in 1930 in the Isle of Wight. Its designer was not Maunsell, but A.B. MacLeod, Assistant for the Isle of Wight, but in practice the de facto General Manager of the island's railway. In 1930, MacLeod had built a 0-4-0 geared hand-shunter named *Midget*. According to P.C. Allen in the *Railway Magazine* for September 1932, the little locomotive with two men working it could haul a fully loaded wagon, three empties or a bogie coach. Michael Robbins in *The Isle of Wight Railways* (Oakwood 1974) said that it lasted until 1938. The *Locomotive Magazine* of January 1932 describes *Midget* thus:

A very interesting little manual rail-tractor may frequently be seen in the works yard (at Ryde). This vehicle is appropriately named *Midget*, runs on four coupled wheels, 1 ft. 2 in. diameter, with a wheelbase of 5 ft., and weighs 1 ton 15 cwt. Designed by Mr. A. B. MacLeod, the Assistant for the Isle of Wight to the Chief Mechanical Engineers', Traffic and Locomotive Running Departments, it was constructed entirely from scrap. The tractor is worked by turning large hand-wheels, which transmit the power direct to a gear-box, giving gears of 1 to 1 and 4 to 1, and thence by chains to the driving wheels. With two men at the wheel, a maximum load of about 20 tons can be moved in low gear. *Midget* is found most useful for moving boilers, carriages, wagons, etc., about the works and yard, and doing work which would otherwise require the services of a locomotive or a gang of men.

The article in which *Midget* is mentioned includes a photograph of the little locomotive.

## ELECTRIC MULTIPLE UNITS

Further 3SUB emus were constructed up to the end of 1930 by placing the bodies of existing pre-grouping suitably modified carriages on new electric underframes and these included those of the former five car units of the 1925 Coulsdon and Sutton ac electrification minus the motor luggage van and one of the trailer coaches, from which the motor luggage vans became bogie goods train brake vans. There were also two classes of 2SUB units, these being converted from the ac South London emus dating from 1909 and Crystal Palace lines emus dating from 1911.

## RAILCAR

The railcar was a four-wheel 50hp petrol railcar built by Drewry and which was ordered the SR in June 1927 and delivered in the following March. The purpose of the SR buying the railcar was an attempt to see if it was possible to reduce expenditure on some of

**The four** wheel 50 horsepower petrol railcar built by Drewry which was ordered in June 1927 and delivered in the following March was the SR's first venture into internal combustion power. In 1934 it was sold to the Weston, Clevedon and Portishead Railway. *Colonel Stephens Railways Museum*

the company's branch lines. The railcar remained with the Southern until it was sold to the Weston, Clevedon and Portishead Railway in 1934. Four similar vehicles were put into service in 1928 to replace the electric trams on Ryde Pier.

## SHIPS

The following ships were built during the period of this chapter.

For the Isle of Wight services, the Caledon Shipbuilding Company of Dundee built the paddle steamers PS *Merstone* and PS *Portsdown* for the Portsmouth to Ryde service in 1928. Fairfield, Govan built in 1930 the paddle steamers PS *Southsea* and PS *Whippingham* for the Portsmouth to Ryde service. The PS *Whippingham,* as well as operating services from Portsmouth to Ryde also operated excursions from Portsmouth. John Samuel White and Company at Cowes built the paddle steamer PS *Freshwater* for the Lymington (Dorset) to Yarmouth (Isles of Wight) service in 1927. William Denny and Brothers of Dumbarton built the MV *Fishbourne* in 1927 and the MV *Wooton* for the Portsmouth Harbour to Fishsbourne car ferry service in 1928.

For the services from Southampton to Le Havre/the Channel Islands/St Malo, William Denny and Brothers of Dumbarton constructed the SS *Isle of Jersey* and SS *Isle of Sark* in early 1929, both entering service in 1930.

For the Newhaven to Dieppe service which was one third owned by the Southern and two thirds owned by the State Railway of France, William Denny and Brothers of Dumbarton built the SS *Worthing* in 1928

For the Dover to Calais and Folkestone to Boulogne services, William Denny and Brothers of Dumbarton built the TSS *Canterbury* in 1928. The ship was built as a first class only passenger ferry.

For cargo services from Dover to Calais and Folkestone to Boulogne, D. and W. Henderson and Company Limited of Glasgow built the twin screw TSS *Deal* in 1928.

# Chapter 3

# THE YEARS OF THE FIRST MAIN LINE ELECTRIFICATIONS

On 7 May 1932, Everard Baring, the Chairman of the SR, died in office. He was succeeded by Gerald Loder, who was created Baron Wakehurst of Ardingly in June 1934, and resigned as Chairman at the end of the year. He was succeeded in 1935 by Robert Holland-Martin.

**This is** an aerial view of Portsmouth and Southsea Station in about 1935 with the line to Portsmouth Harbour crossing the street in the bottom left hand corner. The station opened in June 1847 and was a terminal station until the opening of Portsmouth Harbour Station in October 1876. *Commercial postcard*

## LONDON TRANSPORT

In 1933 the London Passenger Transport Board (LT) was created to operate local transport services within an area whose boundaries in relation to the Southern were, from east to west, Gravesend, Wrotham, Orford, Sevenoaks, East Grinstead, Crawley, Horsham, Ewehurst, Guildford, Woking, Virginia Water and Windsor. This included not only buses, but also trams, trolley buses and trains other than those operated by the four main line railway companies. Unfortunately, the Metropolitan Railway, which regarded itself as a main line railway company, had not been included in the 1923 grouping and despite strenuous opposition was taken over by the Board. Any operators of local services wholly within the Board's area (an approximate radius of 30 miles) were taken over by the Board. LT operated some bus services and a Green Line coach service outside of its area in the area served by the Southern, its terminus being Tunbridge Wells. The reason the four main line companies were excluded was they would not agree to their London suburban services being included in the Board, arguing that they could not be separated from their long-distance networks.

## THE PICCADILLY TUBE EXTENSION

The Piccadilly Line had originated in the 1900s as the Great Northern, Piccadilly and Brompton Railway which was an amalgamation of the Great Northern and Strand and the Brompton and Piccadilly Circus railways and, with the MDR, was part of the Underground Group. The railway had opened from Finsbury Park to Hammersmith on 15 December 1906. On 30 November 1907, a branch line was opened from Holborn to the Strand, later renamed Aldwych. This latter had been planned as part of the Great Northern and Strand prior to the formation of the Great Northern Piccadilly and Brompton and was the only railway branch line in central London. The Great Northern, Piccadilly and Brompton, the Baker Street and Waterloo and the Charing Cross, Euston and Hampstead tube railways had been merged in 1910 as the London Electric Railway, whilst still keeping their identities.

After the First World War there were proposals to extend the Piccadilly tube, both of which came to fruition. One was to extend the line north of Finsbury Park to Cockfosters and this was opened in stages; from Finsbury Park to Arnos Grove on 19 September 1932, to Enfield West (now Oakwood) on 13 March 1933 and to Cockfosters on 31 July 1933

Prior to the First World War, powers were obtained under the London Electric Railway Act 1913 to extend the Piccadilly tube west from Hammersmith to connect to the LSWR's Richmond branch track, but the outbreak of the war put this on hold.

After the First World War, the MDR was becoming crowded. It was running trains from Barking to Ealing, Hounslow, Richmond, Uxbridge and Wimbledon as well as the shuttle service to South Acton and sharing in the Circle line and was a partner with the Midland Railway in the Ealing to Southend through service.

In 1926, powers that existed for the extension of the Piccadilly tube west from Hammersmith were renewed. In 1929, the proposed split of the services west of Hammersmith was clarified. The Piccadilly would run services through to South Harrow and Hounslow, whilst the MDR would run services to Ealing, Wimbledon and Richmond plus the shuttle service to South Acton and the shuttle service from South Harrow over the MR to Uxbridge. There would also be peak hour services to Hounslow.

In 1932, the former LSWR tracks from Turnham Green Junction to Studland Road Junction were transferred to the MDR on a long-term lease and connected with the Piccadilly tube and the former LSWR connection to Kensington (Addison Road – now Olympia) was broken.

On 4 July 1932, Piccadilly trains were extended from Hammersmith to South Harrow with the MDR continuing to run a shuttle service from South Harrow to Uxbridge. On 9 January 1933, Piccadilly trains started to run to Northfields on the Hounslow branch and to

Hounslow on 13 March 1933. Metropolitan District trains continued to Hounslow in the peak hours. On 23 October 1933, following the formation of London Transport Piccadilly, tube trains began running through to Uxbridge and the South Harrow to Uxbridge shuttle service was withdrawn. On 12 September 1932, the MDR lines were extended east to Upminster from Barking over an additional pair of tracks.

## MAIN LINE AND FURTHER SUBURBAN ELECTRIFICATION

The intention of the SR to electrify to Brighton first became known in 1929. The impetus for this was the abolition of Railway Passenger Duty levied on first and second class passengers announced in the Budget of 1929 by the Chancellor of the Exchequer Winston Churchill on the condition that the railway companies spent the capitalised value of the duty on improvement and development schemes in order to relieve unemployment.

The first suggestion that was heard of an extension of the electrification towards Brighton, although not to it, was a statement by Sir Herbert Walker reported in the *Daily Chronicle* of 6 August 1929 when he said that schemes for the extension of the electrification from: Orpington to Sevenoaks; Purley to Three Bridges; Redhill to Guildford; Surbiton to Staines, by way of Weybridge and Virginia Water were being prepared, but that no decision had yet been taken.

According to the *Daily News* of 9 September, the SR proposed to eventually extend its electrification to Chatham and Brighton.

It was on 13 October 1929 that the SR announced its intention to electrify to Brighton. According to the *Daily News* of 14 October:

SOUTHERN'S BIG DEVELOPMENT. The electrification of the London to Brighton main line, which was forecast in the *Daily News* last September, is included in a great improvements scheme involving an expenditure of £2,400,000 announced by the Southern Railway last night. Under the conditions attaching to the recent remission of passenger duty, the Southern Railway guaranteed to undertake an expenditure of 80 per cent, of the capitalised amount (approximately £2,000,000) on improvements to its system. The directors have therefore authorised the following electrification schemes, work on which will begin during the next two or three months The main lines throughout from London to Brighton, a total distance of 51 miles, of which 14 miles to Purley are already electrified. From Preston Park (on the Brighton main line) to Hove and Worthing, 11 miles From Redhill (also on the Brighton main line) to Guildford, via Reigate and Dorking Town, 20 miles. It is hoped that all the work will be completed in two to three years' time. LONGEST ELECTRIFIED ROUTE. When completed, this will be much the longest electrified route in Great Britain, and the first main line to be completely equipped for electric traction. About 1,750,000 passengers are conveyed by rail annually between London and Brighton, and there are about 2,700 season ticket holders in addition. The line from Waterloo to Guildford via Cobham is already electrified. The trains will run on the direct current third-rail system, which is now standard throughout the Southern electrified area. "We have not worked out any details yet so I cannot say whether the journey by electric train will be made in a shorter time than by steam," said Sir Herbert Walker, the Southern's general manager, to the *Daily News* last night.

According to Walker, goods trains would continue to be worked by steam as would the service to Eastbourne. Walker described the scheme as more or less a suburban electrification, which probably would not have pleased the inhabitants of Brighton who regarded the line to London as a main line. Walker said that in addition to the £2,000,000 spent on the electrification £400,000 would be spent on new carriages. This hints at multiple unit electric trains as opposed to locomotive haulage.

According to the *Worthing Herald* of 19 October:

> ELECTRIC TRAINS TO WORTHING IMPROVEMENTS ON THE SOUTHERN RAILWAY THE LONGEST ROUTE . The line from Worthing to London via Hove and Preston Park is included in a scheme for the electrification of a section of the Southern Railway... The cost of the work is estimated at £2,000,000. It is hoped that the scheme will be completed in between two and three years. It will be much the longest electrified route in Great Britain and the first main line to be completely equipped for electric traction. The Mayor of Worthing (Alderman W.T Frost), interviewed by a Herald representative at the beginning the week, said he had received an official communication on the subject from Sir Herbert Walker, the manager of the Southern Railway and he understood the work was to be commenced within three months.

On the other side of the River Thames the inhabitants of the other popular seaside resort for Londoners – Southend – were probably not very happy as they had been promised electric trains before the First World War and still had not got them and there was no sign of them at the time.

According to Michael Bonavia, Walker made his announcement at the Traffic Officers Conference in 1929. 'Gentlemen, I have decided to electrify to Brighton.' Bonavia says that 'a slightly sour reflection might have crossed the mind of any LBSCR man. It would have been that if, three years previously, it had not been decided to convert the AC network to DC, by 1929 the electric trains could have already been running to Brighton.' Bonavia goes on to say that if locomotive traction had been included, the locomotives that hauled the fast passenger trains during the daytime could have hauled the freight trains at night. This would have been an economical arrangement reducing the amount of steam traction.

Even before the announcement of the electrification, it is clear that the SR was looking into the type of electric train for any mainline extensions – electric locomotives or multiple units, as, according to the *West Sussex Gazette* of 19 September 1929, the company apparently had been conducting experiments. Whilst locomotive propulsion would enable existing rolling stock to be used, multiple units would enable a more frequent service. Given that the SR did not own any electric locomotives at the time other than two small shunters the statement is puzzling, unless something was done using the motor luggage vans that were used to propel the ac trains to Coulsdon and Sutton.

In choosing multiple units the SR started a precedent which has become the norm now. The *Daily Chronicle* of 13 January 1930 reported that special coaches would be built for the electrified London to Brighton service. There is no mention of electric locomotives. This clearly indicates the use of electric multiple units and was confirmed at the Southern's annual general meeting on 27 February.

This use of electric multiple units for passenger trains meant that any passenger train from outside the electrified area would have to be hauled by a steam locomotive.

As mentioned, although the SR also originally intended to electrify from Redhill to Guildford and according to Klapper some publicity on those lines appeared 'which included a reference to a loop from London Bridge to Guildford and Waterloo', Walker realised very quickly that it was not the time to reinstate the former SER service that ran from London Bridge via Redhill to Guildford and ultimately Reading. He realised that housing development along the south side of the North Downs was likely to be slow and so only the section from Redhill to Reigate was electrified.

Things started to progress. According to the *Locomotive Magazine* for October 1930, the 'Metropolitan-Vickers Electrical Co. Ltd. have received an order from the Southern Ry. for the equipment of 104 motor coaches and 104 trailers, for use on the forthcoming electrified service to Brighton, as well as for general additions to the rolling stock.' The February 1932 edition of the magazine reported, 'Southern Ry. In connection with the London-Brighton and Worthing electrification, contracts have been placed for forty-four all-steel motor coaches. The order has been divided equally between the Metropolitan-Cammell Carriage, Wagon & Finance Co. Ltd., of Saltley and the Birmingham Railway Carriage & Wagon Co. Ltd., of Smethwick.' But it was not just new trains. There was new signalling as the same edition of the magazine reported:

The Southern Ry. has placed an order with the Westinghouse Brake and Saxby Signal Co. Ltd., of King's Cross, for the supply of material for the new signalling installation at Brighton. The contract comprises electric locking frames, illuminated diagrams, points machines, coloured-light signals, etc. This installation will complete the light signalling system from Coulsdon to Brighton.

New substations for the electrification were required. The *Locomotive Magazine* for July 1932 said:

At Three Bridges the central control station has been erected and equipped to control and operate the current of the eighteen sub-stations. The design of this station excludes all sunlight, and the artificial lighting of the interior has been concealed to avoid interference with the coloured lights on the switchboard. From the main switchboard at this station every switch and circuit breaker on the Coulsdon to Brighton extension, of which there are about 350, can be opened or closed, and its position indicated by a coloured light. Instruments on the board also show the rate of consumption of current in amperes, and the pressure in volts at each sub-station. Apart from an initial movement by the operator, subsequent operations at the station are automatic.

South of Purley the current is being derived from the Central Electricity Board's "grid," and is delivered to the rails at the sub-stations.

The edition for August 1932 reported:

Batteries for the Southern Ry.—All the batteries required for the signalling equipment of the Southern Ry electrification scheme to Three Bridges, which was brought into operation on July 17, have been supplied by the D.P. Battery Co. Ltd., Bakewell, Derbyshire. A battery of 56 cells is to be installed at Haywards Heath on the further section to Brighton, and at Brighton station there will be two batteries, each of 60 cells.

The period of the electrification to Brighton saw the end of slip coach working on the SR. The last slip coaches were the Horley slip running through to Forest Row off the 5.05 pm express from London Bridge to Eastbourne and the Haywards Heath slip which was the slow portion for Brighton off the 5.08 pm London Bridge to Worthing and Angmering and also on Saturdays only the Three Bridges slip off the 5.20 pm; Victoria to Eastbourne. The last slips taking place on Saturday 30 April and in each case a stop replaced the slip. So that the Horley stop of the 5.05 pm from London Bridge did not delay the 5.08 pm from London Bridge the two trains were transposed. The 5.08 left at 5.04 and the 5.05 at 5.08 and their overall times increased by 4 minutes. .

17 July 1932 saw the first stage of the electrification when electric trains started running from Coulsdon to Reigate and Three Bridges, via Redhill. For this thirty-five new four-car emus were provided. These trains were made up of two motor third class brakes, one lavatory composite, and one non-lavatory composite, the seating accommodation of each unit being seventy first-class and 204 third-class.

For the full service to Brighton including the all-Pullman *Southern Belle* train, new express trains were needed. For these, the following trains were built: twenty six-car emus, gangwayed within the set and including a Pullman car, three similar units, but with rather more first class accommodation used on the *City Limited* between Brighton and London Bridge and return and three five-car all Pullman trains. The *Locomotive Magazine* of February 1933 specifically mentions Pullman cars, saying:

New Pullman and Motor Cars for the Brighton Electrification Extension of the Southern Ry.—The thirty-eight new Pullman cars to run in connection with the Brighton and Worthing electrification, as well as twenty-two of the motor coaches, were built by the Metropolitan-Cammell Carriage, Wagon and Finance Co. Ltd., at the Midland and Saltley Works. The Pullman cars were designed in collaboration with Mr. W.J. Sedcole, Engineer of the Pullman Car Co., while the motor coaches were built to the requirements of Mr. R.E.L. Maunsell, CBE, the chief mechanical engineer of the Southern Ry.

According to the *Railway Magazine* for December 1932, the electric trains first ran on test to Brighton on 1 November 1932 and that the automatic signalling had been brought into use between Coulsdon and Haywards Heath on 3 October 1932, and Haywards Heath and Preston Park on 6 October 1932. The new signal box in Brighton had been brought into use on 16 October 1932 which dispensed with six signal boxes. It added the off-peak service would comprise five trains per hour – one non-stop, two semi-fast, one stopping at intermediate stations to Brighton and one to West Worthing and that additional trains would be run in mornings and evenings. G.T. Moody in *Southern Electric* (Ian Allan 1968) said that the trials started on 2 November and that the trials were conducted using a specially made up five-car train comprised of two motor brake thirds, two ex-LSWR brake thirds and one standard SR third. These together with the four-car emus for the stopping services had been trialled on the Waterloo to Guildford line, but unlike the latter had not been used in passenger service. After the trials the motor coaches of the five-car unit were used on emus for the *City Limited*.

According to the *Locomotive Magazine* for January 1933, prior to the inauguration of the electric services:

during the week following Christmas an exhibition of the new main-line electric trains took place at Victoria Station, London, No. 17 platform, and at Brighton, No. 7 platform. At Victoria a six-coach corridor train, and the new electric *"Southern Belle"* train, were shown. At Brighton there was a new six-coach electric train, a full-size model of Stephenson's Rocket, Stroudley's 0-4-2 express engine of 1891, No. 2172 (originally No. 172 *Littlehampton*), and No. 853 *Lord Rodney*, four-cylinder express locomotive. No. 2172 is the last of the famous "Gladstone" class to remain in service … As from January 1 the whole of the ordinary passenger services of the Southern Ry. between London, Brighton, and Worthing have been electrically operated. The distance from London to West Worthing is 61¾ miles, and this constitutes the longest continuous electrified route in this country.' and 'Recently, with steam operation, twenty-seven trains made the run in both directions. The new time-table shows fifty-three electric trains leaving London for Brighton each week-day and fifty-one trains from Brighton to London. The best time between London and Brighton is still one hour, which has been done for years, whilst 1 hour 35 minutes to 1 hour 40 minutes for stopping trains is well within the capacity of the steam locomotive…..On Friday, December 30, the Lord Mayor of London visited the Mayors of Brighton and Worthing in a special train run for the private inauguration of the new service.

The *Daily Mirror* 31 December 1932 carried an article headed:

ALL-ELECTRIC LUXURY EXPRESS. Run to Brighton Opens New Coast Service. WAVING CROWDS' and which began 'BY A SPECIAL CORRESPONDENT With the Lord Mayor of London (Sir Percy Greenaway), the Sheriffs, and many other distinguished personages, I made the first run to Brighton yesterday in one of the new all-electric Pullman trains. It was the most luxurious and enjoyable railway trip I have experienced. Sitting back contentedly on the comfortable cushioned seats, we were whirled away from Victoria with incredible smoothness at a speed which would get us to Brighton within the hour. The designers of these trains have considered every detail which will add to the comfort of passengers.

The *Worthing Gazette* of Wednesday 4 January 1933 reported.

INAUGURATION OF THE NEW ELECTRIC TRAIN SERVICE. HOW WORTHING OBSERVED THE OCCASION. THE LORD MAYOR AND SHERIFFS WELCOMED. RAILWAY'S PLEA FOR FAIR TREATMENT. A new era in modern travel was successfully inaugurated on Friday when. at the invitation of the Southern Railway, the Lord Mayor of London (Sir Percy Greenaway) and the Sheriffs of the City travelled to Worthing and thence to Brighton in one of the new electric express trains, officially to inaugurate the electrified extension of the main line from London to Brighton and Worthing. The electrification scheme, including the provision of new rolling stork and equipment, has cost £2.750,000, and it is confidently anticipated that it will open up a new era of prosperity for the coastal towns served by it, inducing many more London business men to become permanent residents. Incidentally, in a speech which he had prepared for the occasion but which in his absence owing to illness, had to be read for him, Sir Herbert Walker, the General Manager of the Southern Railway, made the interesting announcement that, in view of the outstanding success already attending electrification in the suburbs and on the main line to Three Bridges, the Company were turning their minds to the examining of further electrification schemes.

It should not be thought that the end of steam did not see some commemoration.

According to Railway Correspondence and Travel Society's magazine the *Railway Observer* for February 1933, on the last day of 1932 the devotees of the old LBSCR were out in force both in London and Brighton to see the last of the steam service and many of them made the journey to Brighton and back. A number of photographers were busy although the day was not ideal for photography. On that day, which was a Saturday, the Down morning *Southern Belle* was hauled by King Arthur Class 4-6-0 No 795 *Sir Dinadin* whilst on the return journey the locomotive was No 804 *Sir Cador of Cornwall* of the same class. Both the up and down afternoon *Southern Belle*'s were worked by former LBSCR Class L 4-6-4 side tank No 2333 *Remembrance*. The last steam train of all was the 12.05 am Victoria to Brighton on 1 January 1933 and which was hauled by Class L No 2329 *Stephenson*. At 11.50 pm, about twenty railway enthusiasts chatted on Platform 16 with Driver Rogers and Fireman Stoner of the locomotive and Inspector Ewres who was travelling on the footplate to Brighton on the last steam journey. At three minutes past midnight H.J. Stretton-Ward the President of the Railway Correspondence and Travel Society climbed up and put a large black bow on the locomotive's smokebox. The engine crew and the inspector got on the locomotive. At five minutes past midnight the guard turned his lamp to green and swung it high. Driver Rogers looked at the signal which was clear for the last steam train to Brighton. The driver and fireman waved their caps. The small group on the platform waved back. The Pullman car attendants and the guard waved back. Three resounding cheers crashed and re-echoed under the station roof. The last pall of steam swirled and tried to hide the train's tail lamp.

The *Weekly Dispatch* of 1 January carried a very brief article which read 'THE LAST BOW – With a large black bow tied to the front of the engine the last steam train for Brighton left Victoria Station this morning at 12,5. In future the whole of this service will be by electric trains.'

This was the last steam train between London and Brighton under normal conditions. In the severe winters of 1940 and 1946-7 and during the ASLEF strike of 1955 there were some steam worked passenger trains between London and Brighton and reverse. There were also some steam worked enthusiasts' specials.

The *News Chronicle* of 2 January 1933 carried a report by H. de Winton Wigley of catching the first electric train in public service from London to Brighton – the 6.44 am from Victoria. Wigley says that he saw the new year in and left home at 5 am to catch the aforementioned train. At Victoria he asked the ticket clerk if he was the first electric return to Brighton that year. The ticket clerk replied that he was. Having got his ticket, Wigley then got the ticket clerk to sign his autograph book. At the ticket barrier, Wigley says that the porter shouted to another porter, called Alf, that here was the first electric return to Brighton and did he want to clip it, which Alf duly did, In his report he mentions a young railway enthusiast travelling on the train. The young man wrote down the number of each carriage and asked for the driver and guard's full names. Clearly not all railway enthusiasts were like those who had gathered at Victoria to see the last steam train out. From Wigley's report it is clear there was a certain degree of informal ceremony in connection with the first public electric train to Brighton. He also returned to London and then took the first electric *Southern Belle* to Brighton. The SR had the London Madrigal Society sing on it. According to the *Daily Herald* of 2 January, the singers sang madrigals and part songs,

**The London** to Brighton and Worthing electrification, which was inaugurated on 1 January 1933 was the first main line electrification in Britain. In this photograph is one of the 6PUL six car emus built for express services on the line. SR Official.

the purpose being to demonstrate the silence and smooth running of the new electric service. According to the *News Chronicle,* the young railway enthusiast was at Brighton to meet the first electric *Southern Belle* in and to collect the first menu card autographed by the Chief Steward.

In 1934, the *Southern Belle* was renamed the *Brighton Belle.* According to the *Locomotive Magazine* of January 1934:

The *"Southern Belle"* Pullman train is to be known in future as the *"Brighton Belle.'* The renaming took place at Brighton Station on 29 June 1934 upon the arrival of the mid-day Victoria to Brighton service when the Mayor of Brighton, Margaret Hardy, renamed the train. The *News Chronicle* of 30 June reported' BRIGHTON'S TWO RECORDS Brighton had two "world record" ceremonies yesterday. One was the renaming of the *Southern Belle*—now the only electric Pullman train in the world—as the *Brighton Belle*. The other was the opening of the new £100,000 swimming stadium—the largest covered sea-water baths in the world.

It was not only the Brighton line that saw electric trains in 1933. On 16 July 1933, electric trains started running over the Lewisham to Hither Green section of the Lewisham loops.

1 May 1934 saw electric trains extended from Bickley to St Mary Cray .The *News Chronicle* of 2 May reported, 'New Link.—The new electric extension of the Southern Railway from Bickley to St. Mary Cray, Kent, was opened yesterday.' According to the edition of 8 May this was all to meet the wishes of the City workers living on the new Orpington Garden Estate. According to Alan A. Jackson in *Semi-Detached London* (Wild Swan 1991) the Garden Estate or 'Orpington Garden Village' as Jackson says it was called, was built by O'Sullivan (Kenley) Ltd who had conveyed to the Southern free of charge land for the enlargement of St Mary Cray station in 1934. According to the *Railway Observer* of May 1934 this was the first part of the Sevenoaks electrified lines.

The decision to electrify to Sevenoaks was taken at the very end of November 1933. *The Bromley and West Kent Mercury* of Friday 1 December 1933 reported, 'ELECTRIC TRAINS TO SEVENOAKS. The electrification of the Southern Railway from Orpington, and also on the St. Mary Cray section, is shortly to take place as far as Sevenoaks. Services to St. Mary Cray will be doubled and those to some other stations trebled.'

The *Railway Magazine* of January 1934 reported that during the previous month the SR had announced its decision to electrify two important groups of lines. These were firstly the two routes to Sevenoaks from Chislehurst and Bickley through Swanley Junction and from Orpington and secondly from Brighton and Wivelsfield to Eastbourne, Hastings and Ore, as well as from Haywards Heath to Horsted Keynes and from Lewes to Seaford.

The *Locomotive Magazine* of January 1934 said:

Southern Railway.— Following closely on their decision to extend the electrically operated portion of their system from Chislehurst and Bickley to Sevenoaks, via Swanley and from Orpington to Sevenoaks via Chelsfield, it was announced at the beginning of December that it had been decided to extend the main line electrification scheme from Wivelsfield Junction to Eastbourne and Hastings and from Haywards Heath to Horsted Keynes at a cost of £1,750,000. The extension is expected to be completed early in 1935. A quantity of new rolling stock will be required which will include seventeen six-car units, five four-car units, and eighteen three-car units with a number of Pullmans. Platform extensions will be necessary at London Bridge, Lewes and Eastbourne and alterations at Bexhill Central.

Of the two schemes, that to Eastbourne and Hastings had first been proposed by Walker in May 1932 – even before the completion of the electrification to Brighton. According to Klapper, when Walker had first submitted the proposal to electrify from Keymer Junction and Brighton to Lewes, the Seaford branch and along the coast to Eastbourne and from Polegate to Hastings the SR's board was somewhat less than enthusiastic. Klapper says that the board although no doubt good businessmen did not see as far ahead or as optimistically as Walker. The overall cost for the scheme was £1,787,750. About half of this was to come from depreciation funds. The mileage for conversion was 60½ miles and it was forecast that there would be a reduction in the cost of operation. On 29 June 1933 a revised scheme was put forward. The proposal now was for 62 route miles and 129 track miles at a total cost of £1,740,511. Of this, only £962,451 was to be charged to the capital account. Klapper says that with the Brighton line figures in front of the board it might have been easy, but it was deferred first to the July board meeting and then to the meeting of November. It was not until the board meeting of 30 November that Walker was able to persuade the Southern's board to approve the electrification from Brighton to Lewes, Wivelsfield to Eastbourne, Hastings and Ore, as well as from Haywards Heath to Horsted Keynes and from Lewes to Seaford.

On 6 January 1935, electric trains were extended to Sevenoaks (Tubs Hill) from Orpington and from Bickley and Chislehurst and St Mary Cray via Swanley and Otford. *The Sevenoaks Chronicle and Kentish Advertiser* of Friday 4 January 1935 reported:

SEVENOAKS AS OUTSIDE RIM OF LONDON. WHAT RAILWAY ELECTRIFICATION MEANS. NEW RESIDENTS WILL BE WELCOMED. On Sunday next January 6 electric trains will come into operation between Sevenoaks and London and 23 more route miles will added to the Southern electric system The cost of the work, which has meant considerable additional employment. has been over £500.000 and instead of 31 trains a day from London and 27 to London there will 63 'down' and 48 'up' trains from Tubs Hill. At Otford the increase is even greater, with total 113 trains a day against the present total of 42. These figures include trains going in both directions. The electric trains will stop at all stations on the newly electrified extensions The service between Sevenoaks and Charing Cross or Cannon-street via Orpington will be supplementary to the steam trains and will consist of three trains per hour at peak hours in week-days and two trains per hour during the slack periods.

The Brighton to Lewes, Wivelsfield to Eastbourne, Hastings and Ore, Haywards Heath to Horsted Keynes and Lewes to Seaford electrification left the L class 4-6-4 side tanks without suitable occupation owing to their limited water and coal capacity and according to the *Railway Magazine* of July 1934 it was intended to convert them to 4-6-0 tender locomotives to rank in power with the King Arthur Class 4-6-0s.

Rolling stock for the new service comprised of express trains of some six-car emus with a pantry car instead of a Pullman car in the set whilst for stopping services there were some two-car non-gangwayed units having a lavatory in each carriage and for local services some two-car non gangwayed units not having a lavatory in either carriage.

On certain sections of line where there were metal sleepers these had to be replaced with wooden ones.

Trial running started in May 1935 with the formal opening taking place on 4 July and public services commencing on 7 July.

On 4 July, the Lord Mayor London, Sir Stephen Killik and the Sheriffs of London accompanied by the Chairman, Directors and Officers of the Southern travelled by special train to Eastbourne, Bexhill and Hastings to meet the mayors of those resorts. According to the *Hastings and St Leonards Observer* of Saturday 6 July 1935:

The journey of the first official train from London to Hastings, via Lewes, Eastbourne and Bexhill, was indeed a triumphant one. Few trains can have been so much photographed as this luxurious twelve-coach pioneer of a new era. All along the line amateur photographers waited, with cameras poised, for the passing of the train. Its journey through Sussex was watched by countless faces, nearly every window of houses overlooking the line having its complement, and hundreds of children cheered it on its way. The ceremonies at Eastbourne, Bexhill and Hastings were watched by a large number of people, and the stations and nearby streets were decorated with flags and bunting. On arrival at Eastbourne Station, the Lord Mayor and Lady Mayoress (Mrs. Stanley Greenland), who were accompanied by the Sheriffs, Aldermen and Councillors of the City of London, and the chairman, directors and other chief officers of the Southern Railway, were received by the Mayor of Eastbourne (Miss E.M. Thornton. JP), the Mayoress (Miss Madge Thornton). Mr. Charles Taylor (MP for Eastbourne), Colonel R.V. Gwynne (Deputy-Lieutenant of the County), and other Town Councillors and officials. Also on the platform was 99-year-old Mrs. C. Allchorn, who watched the first steam train puff into Eastbourne 80 years ago. The Mayor of Eastbourne thanked the Lord Mayor for the interest he displayed in the South Coast towns, and in reply the Lord Mayor said he was delighted to have the opportunity of renewing his acquaintance with their attractive town. He added that the electrification of the railway would have the effect of bringing Eastbourne still nearer to the capital, and would mean increased prosperity both to the town and to the railway. Luncheon followed at the Grand Hotel, the Mayor of Eastbourne presiding. In the evening there was a banquet at the Queen's Hotel in Hastings presided over by the Mayor of Hastings.

The *Hastings and St Leonards Observer* reported that the Mayor said, 'it is a great day for Hastings and St. Leonards. Your visit, my Lord Mayor and Lady Mayoress, with your Sheriffs, will be remembered by us with pride for many years to come.' There were some comments at the dinner on the fact that the electric service to and from Hastings was by the long route (via Eastbourne) and not the short route (via Tunbridge Wells), and these drew from Mr. R. Holland Martin (Chairman of the Southern Railway) the following statement. 'The direct line is a very difficult one, a line which has tunnels rather too small to have special stock run through them. But you are not going to get a worse service, you are going to get a better and better service on that line and on the electric line we have given you to-day.'

According to the *Daily Mirror* of 5 July, the inaugural train was driven by James Arthur Allen, who had been on the railway since 1908 and had given up a day of his holiday to do so. He was a member of Eastbourne Borough Council and it was the first occasion that he had been in charge of a main line electric train. The reason that he was driving was at the suggestion of the Mayor of Eastbourne, Miss E.M. Thornton. Unfortunately, he did not attend the civic lunch, as he was at home having his lunch and later returned to drive the guests to Bexhill where they had tea and then to Hastings for the banquet. One has to say that given that he had given up a day of his holiday that it was a bit tough on him and that he should have been invited to the civic lunch! According to the *Daily Herald* of the same day, Allen was a Labour councillor.

Owing to the regulations regarding the working of vans in multiple unit trains the Newhaven boat train service remained steam hauled.

To complete the electrification in the suburban area the Nunhead to Lewisham section of the Lewisham loops was opened for passenger trains – both steam and electric on 30 September. On the same day the former LBSCR and SER jointly owned Woodside and South Croydon Railway from Woodside to Selsdon Road, which had closed to passenger traffic other than the odd through excursion on 1 January 1917 was reopened and electric trains extended via the line to Sanderstead.

According to Charles Klapper, before the Second World War the Southern put up an electrification scheme for the West London Extension Railway only to be met with objections from the GWR and the LMS that three electric trains per hour would interfere with goods services even though the LMS ran goods trains to Norwood Yards on the Southern between nine electric trains per hour. As to what year this was is not clear. It is worth pointing out that electric trains were run by the LMS over the West London Railway from Willesden Junction to Earls Court and the MR, later LT over part of the same railway from Edgware Road to Kensington (Addison Road). The

West London Railway being jointly owned by the GWR and the LMS. The MR connection being that trains on the service to Kensington were Hammersmith and City trains and the Hammersmith and City Railway was jointly owned by the GWR and the MR.

## OPENINGS OF LINES

There was only one new railway opened during the period covered by this chapter and that was the 1¾ mile long branch from Stoke Junction on the Port Victoria branch on the Isle of Grain to Allhallows on Sea. There is some confusion as to the actual date of opening. Dendy Marshall says 14 May, the *Daily Herald* of 11 May 1932 says Whit Monday, which was 16 May and the *Sheerness Guardian and East Kent Advertiser* of 14 May says 17 May. The *Railway Observer* for June 1932 carried a report by A.W. Reed who said that he travelled on the branch on the opening day 15 May. I am going to stick with 15 May given that it is contemporary.

**The Stoke** Junction to Allhallows-on-Sea branch opened in mid-May 1932 and in this photograph we see the terminus of the line at Allhallows-on-Sea prior to opening but with the track fully laid. The contractor's locomotive is a Manning-Wardle saddle tank. *SR Official*

The background to the short branch was that in the Hundred of Hoo, which is a rather desolate peninsula thrusting out between the rivers Thames and Medway, there was a small village called Hoo Allhallows. This lay on somewhat higher ground than the surrounding marshes which formed a large part of the area. On the slope between the village and shore a land development company called the Allhallows-on-Sea Estate Limited had laid out quite an extensive estate and a hotel. Geographically, Allhallows is opposite Leigh-on-Sea in Essex, which since 1913 had been part of Southend. The idea appears to have been to try to rival Southend. Sadly this did not happen.

In March 1929, Walker reported to the SR board on the proposed development at Allhallows. According to him, a relatively short piece of railway could serve the development and would only cost £56,000, but Basil Scruby and Company acting for the owners of the estate offered to contribute £20,000 to the sum and give the land free of charge where it would pass through their property. The SR wasted no time and at the meeting of the Company's board in May it was reported that an application for the Allhallows Light Railway had been made to the Ministry of Transport. In June 1929 the Light Railway Order was granted.

In 1934, Allhallows-on-Sea Estate Limited sold its property to Aynsley Trust Limited and at the SR's board meeting in December that year it was reported that the company had a capital of £250,000 and it was proposed that the SR should take £50,000 in shares, £18,000 of which would be in satisfaction of debt.

## CLOSURES OF LINES

The years 1931 to 1935 saw quite a few closures. As mentioned, the Canterbury and Whitstable line closed to passenger traffic from 1 January 1931. The same day saw the closure to passengers of the branch line from Fort Brockenhurst to Lee on Solent with a replacement bus service being provided and complete closure on 30 September 1935.

On 1 April the section of line from Hythe to Sandgate was closed and the passenger service replaced by buses.

On 6 July 1931 passenger services ceased on the line from Hurstbourne Junction to Fullerton Junction and on 29 May 1934 the Junction near Hurstbourne was removed, followed on in October 1934 with the removal of the track from Hurstbourne Junction to Longparish. Goods traffic continued to be worked from Fullerton to Longparish.

At the end of April 1932, the SR announced its intention to close the Basingstoke and Alton Light Railway, which it had been forced to re-open in August 1924. According to the *Hampshire Telegraph* of 29 April, Alton Urban District Council and Alton Rural District Council were not taking any action, but at Basingstoke Town Council:

Councillor Dellafera said the County Council had just incurred a very big outlay in constructing a new bridge over the line in connexion with the new by-pass road. He considered that the Railway Company should have given the 'County Council an opportunity of avoiding that expense to the ratepayers. The matter was referred to the Finance end General Purposes Committee.

Passenger services were withdrawn from 12 September. According to the *Hampshire Telegraph* of 9 September 1932:

LIGHT RAILWAY TO CLOSE Basingstoke-Alton Line TRAINS TO RUN FOR FREIGHT TRAFFIC The Southern Railway announce that on and from Monday, September 12. the services of passenger trains on the Basingstoke and Alton Light Railway will be withdrawn, although the line will continue to be worked daily for freight traffic. The stations affected are Cliddeaden, Herriard, and Bentworth. Milk traffic will continue to be conveyed, and after the withdrawal of the train service passengers will receive both morning and afternoon facilities equal to those at present in present in operation between Herriard Station and Basingstoke by a service of road motor vehicles based on Basingstoke.

Goods services survived a few more years until 1 June 1936. According to the *Hampshire Telegraph* of 5 June 1936:

> The Alton-Basingstoke light railway was finally closed on Saturday, and no longer will a train puff its way through the winding, serpentine stretch which separates the 13 miles between the towns. The line closed for passenger traffic two or three years back, and the track from Alton to Bentworth and Lasham was pulled up, the remainder being kept open for goods traffic between Bentworth and Basingstoke.

As to when the track from Alton to Bentworth and Lasham was pulled up is not clear, but contemporary newspaper reports suggest the latter part of 1932.

2 January 1933 saw three closures. In the Brighton area the branch line to Kemp Town lost its passenger service, but that to the Dyke lost its goods service and did not lose its passenger service until 1 January 1939. The branch line from Botley on the Fareham to Eastleigh line to Bishop's Waltham was the other branch line to lose its passenger service on 2 January 1933.

On 17 June 1933 the section of the SDR from Corfe Mullen Junction to Wimborne lost its goods service. Its passenger service had been withdrawn before the grouping.

Finally for 1933, the goods only branch to Ruthern Bridge in Cornwall was closed on 30 December, although the last train had run on 29 November.

1935 saw the closure of three lines and one dependent railway. The first line to close was that from Chichester to Midhurst on 8

July 1935, but only for passengers. The *Hampshire Telegraph* of Friday 12 July 1935 reported 'MIDHURST CHICHESTER LINE CLOSED – On Saturday the railway line from Midhurst to Chichester was closed to passenger traffic. The line was opened on July 11, 1881, and one of the passengers on the last train was George Batchelor, who was the fireman on the first train.'

Of the two lines to close on 29 September, one was the section from Ringwood to Christchurch and a replacement bus service was put on for passengers. According to the *New Milton Advertiser* the last train left Ringwood to the accompaniment of cheers and fireworks and at Hurn some people gathered to sing 'Auld Lang Syne'. The line was closed completely.

The other line to close was the Lynton and Barnstaple. The announcement of the closure was made quite early in the year. Press reports start mentioning it as early as mid-February. For example according to the *North Devon Journal* of Thursday 21 February 1935:

SUGGESTED CLOSING OF LYNTON RAILWAY. Special Meeting of U.D.C. We were informed yesterday that notices have been sent out to members of Lynton Urban District Council calling a special meeting for to-morrow (Friday) "to consider the decision of the Southern Railway to close the Lynton line at the end of the summer season." The way in which the agenda for the meeting is worded suggests that the Council have received some official intimation to justify a special meeting being summoned. Officials at the London headquarters of the Southern Railway, however, decline to confirm or otherwise the rumour, that has been an open secret for over a week.

Following a recent report that the Lynton and Barnstaple Railway may possibly be closed at the end of next summer, a deal of controversy is taking place. Such an action would greatly affect Lynton, Lynmouth, and neighbourhood. So many visitors speak most highly of the quaintness of the railway and the charming scenery through which it runs. The importance of the railway to the seaside resort cannot be over-estimated.

Protests were made to the SR, which argued that the line was running at a loss, which were of no avail. According to the *Railway Observer* of March 1935, a letter from the SR stating that they were closing the narrow-gauge railway between Barnstaple and Lynton at the end of the season aroused protest at a meeting of Lynton Urban Council. A resolution was passed unanimously urging the Southern not to close the line.

The *Shepton Mallet Journal* of Friday 29 March 1935 gave the Southern's response as to why the line was going to close. It said:

TOY RAILWAY'S FATE The Southern Railway Company, replying to protests against their proposal to close the Lynton and Barnstaple toy railway, explain that for a number of years they have had to bear a considerable annual loss on the line. "We are now faced with the renewal of the permanent way," they write, and in view of the loss and the apparent impossibility of the line becoming a paying proposition, we have no alternative but to close it." The letter ends "If of course, the various authorities interested would be willing to give us some guarantee to cover the deficit we should be prepared to reconsider the matter."

According to the *Railway Observer* of May 1935 the SR had definitely decided to close the line after the summer season. The *Observer* said that according to newspaper reports, Lynton and Lynmouth sent a delegation to the Southern's representatives in Barnstaple to urge its continued operation and adds that all but one of the delegation which went to see the SR at Barnstaple went by car.

This was not the only problem. The *Western Morning News* of Monday 20 May 1935 reported that on the first long day Sunday excursion of the summer season, from Barnstaple to Lynton, leaving Barnstaple at 11.50 and returning from Lynton at 8.10 the number of passengers was six and one half-fare. It is true it was not quite summer weather, but it was an attractively low fare, and the company had advertised and well announced the excursion in the usual way. The gross revenue the railway company gets from these fares is 9s. 9d. and at a modest estimate the running expenses to the company would be £10. This instance is an interesting commentary on the efforts being made in North Devon to induce the Southern Railway to reconsider their decision to close this line at the end of September.

According to the *Railway Magazine* of May 1935, 'Despite local requests the Southern Railway does not find that it can continue to work the narrow gauge line between Lynton and Barnstaple, Devon. It is therefore announced officially that this line will be closed at the end of the summer train service.' The same magazine for November 1935 had a lengthy illustrated article on the line or rather railway by John W. Dorling BSc.

The *Locomotive Magazine* for April 1935 reported 'Southern Railway.— It is announced that the narrow gauge Lynton and Barnstaple line will be closed down on the cessation of the summer train service this year.' In the magazine for September 1935, the railway photographer H.C. Casserley wrote a descriptive and illustrated article on the line whilst the October issue reported 'Southern Railway.—The Lynton and Barnstaple narrow gauge line was closed for all traffic on Sept. 29. The Christchurch and Ringwood section was also shut down for passenger traffic at the end of September.'

According to the *Railway Observer* of June 1935, because the line was closed on Sundays [except for summer Sunday excursions] omnibus connections were provided from Barnstaple Junction and that the time taken by bus to Lynton was 95 minutes compared to the best timing of 84 minutes by rail.

In the week of 22 to 29 September 1935, Olivia Hamilton Ellis, who was the wife of the English railway painter and author Cuthbert Hamilton Ellis, expressed her amazement that he had not seen the Lynton and Barnstaple line. Cuthbert Hamilton Ellis went to Waterloo station to inquire about travelling to the Lynton and Barnstaple line and according to his account published in *Trains Illustrated* in 1955, the staff at the inquiry office there looked bored when he told them that he wanted to make the trip. Whilst there was a night train they did not encourage passengers and technically it did not become a passenger train until it reached Salisbury, so cheap day trips were out of the question! Hamilton-Ellis went off to Paddington, the London terminus of the rival GWR, where a very helpful lady inquiries clerk told him that he must see the Lynton and Barnstaple. Whilst the Great Western did not have an excursion going to Barnstaple, they did have one going to Plymouth leaving at midnight. She told him to get a special cheap-day ticket to Taunton, and when they got there to get a market-day cheap return to Barnstaple. Hamilton-Ellis describes how the two intrepid travellers were at Paddington late that night at the start of their journey. In the article, he described their experiences in getting to Barnstaple and their journey over the Lynton and Barnstaple, their visit to the twin towns of Lynton and Lynmouth and their return journey.

For a railway company that owned a line to actively discourage two travellers from visiting it and its rival company encouraging them to visit it says something.

The *Daily Mirror* of 30 September 1935 carried an article on the closure of the line. It read:

The Last Run of the 7.55 from Lynton. BRASS BAND'S SWAN SONG. BY A SPECIAL CORRESPONDENT WITH a brass band on board to give full musical honours to the historic occasion, the last train ever to run on the Lynton-Barnstaple light railway left Lynton last night carrying 310 local excursionists back to

Barnstaple. Two baby tank-locomotives were required to pull the nine coaches on the memorable last run of the 7.55. .The final chuffs of the two engines were accompanied by tuneful excerpts rendered by the Lynton Brass Band. At Blackmoor—a half-way house on the line—the band left the train to return to Lynton. They made the journey by motor-bus. The last train had gone—for ever. One of the people who most regrets the passing of the famous little line is Station-master F. E. Liley, of Lynton. "When I go back to the main line to my old job," he told me, "Lynton will be left with a disused station—nineteen miles from the nearest railway, which is at Barnstaple. "It is a pity, but the line doesn't pay for itself," In a fortnight's time five engines, rolling stock, rails and sleepers will be put up for auction at Barnstaple. It is expected that they will fall under the hammer to a model-railway enthusiast who wants to try his hand at something nearer the real thing in railways.

According to the *Western Morning News* of 30 September

The Lynton and Barnstaple Railway, opened in 1898, closed yesterday when the last passenger train ran over the line from Barnstaple to Lynton and returned last evening. The two-engined train from Barnstaple carrying 300 excursionists steamed out in sunshine, with whistles sounding defiantly for a distance of more than a mile, and several hundred people had gathered on the North Walk at Barnstaple to wave cheery farewell.

On the return journey Lynton and Lynmouth band played Auld Lyng Syne as the train left Lynton for the last time with whistles shrieking, detonators exploding and the Last Post being sounded. When the train reach Barnstaple at 9 40 pm to the accompaniment of a fusillade of detonators it was greeted by a crowd of about 1,000 people.

The sale of the effects of the Lynton and Barnstaple took place on 13 November 1935 at the railway's Pilton depot in Barnstaple. The dismantling of the railway took until July 1936. Locomotive No 188 was used in this work by Sidney Castle, the dismantler of the railway. The locomotive No 188 was then sold to a coffee plantation in Brazil and apparently lasted until 1957. Apart from three carriages all the other rolling stock was scrapped.

The debate will always go on as to whether the Lynton and Barnstaple could have been saved. It has been said the line could have been operated like the GWR's Vale of Rheidol line. But that misses the point. Devil's Bridge is a tourist attraction. The places served by the Lynton and Barnstaple were working communities which needed a service all the year round, unlike the Welsh line which only ran in the summer after the end of the 1931 summer season. The railway suffered two disadvantages. It was narrow gauge which did not permit through running from the rest of the SR's system and Lynton and Lynton Station was inconveniently situated for the town, whereas buses ran right into the town centre.

## THE WEST SUSSEX LIGHT RAILWAY OR SELSEY TRAMWAY

The other railway to close was the Selsey Tramway or Hundred of Manhood and Selsey Tramway or West Sussex Light Railway running from Chichester to Selsey which closed on 19 January 1935. This was a standard gauge independent railway and totally dependent for through traffic on the Southern, through traffic being in the form of goods traffic. The railway was one of Colonel Stephens' light railways.

The railway had opened on 27 August 1897 from Chichester to Selsey Town and was extended to Selsey Beach in 1898; all without the benefit of legislative sanction. The extension to Selsey Beach closed in 1908.

In 1913, the directors of the Selsey proposed to construct a light railway extension from Hunston to West Itchenor and East

Wittering. The Light Railway Order granting this would also regularise the legal status of the existing line, which was to be called the West Sussex Railway and to improve the line. The Light Railway Order was agreed but the outbreak of the Great War delayed its granting until 1915 and in consequence no action could be taken due to the war.

After the end of the war, new steps were taken to give the Tramway some legal status. A novel legal certificate dated 1924 authorised a new West Sussex Railway Company to take over and reconstruct the line. The proposed extensions to Patcham and Bognor, and also Wittering was pursued but a Light Railway Order for these was never issued. The new Company was also empowered to enter into agreement with the SR with regard to the re-construction, working and management of the line.

As mentioned in the previous chapter, those behind what became the RHDR thought of taking it over and converting it from 4 ft 8½in gauge to 15in gauge.

With the death of Stephens in 1931, the Company, facing increased competition from improved buses, seemed to lose heart. His successor W H Austen made an attempt to get the SR to take over and reconstruct the line but after a detailed survey they declined to do so. The directors therefore decided to close the line which was effected on 19 January 1935.

According to Klapper, prior to this, Walker had had investigations made as to the prospects for electric traction to Hayling Island and Selsey. This was before the proposal to electrify the former LBSCR route to Portsmouth was put before the SR's board. He says that Owen Walker, the receiver and manager of the Selsey Company, offered the Southern the freehold of the company, which had decided to go into liquidation. At the board meeting of the Southern on 26 July 1934, Sir Herbert Walker was authorised to negotiate with Owen Walker. Apparently, Sir Herbert Walker's aide Leslie did a walkover of the Selsey line, but the board was only prepared to offer

£5,000 for the company and in view of the complications of bringing it up to the Southern's minimum standards, nothing more was done in the way of a better offer. In 1934, the SR held 33 per cent of the shares of Southdown Motor Services so that one third of the profit of the bus service from Chichester to Selsey went to the SR anyway. As regards Hayling Island, a new swing bridge to take electric trains would have been required. Southdown also served Hayling Island.

The *Chichester Observer* of Wednesday 2 January 1935 said, 'Selsey. Tramway. The owners of the land through which the Tramway runs are anxiously awaiting the decision of the Southern Railway not to buy the line. In that case, if the Tram Company shuts down, the land of the line becomes, automatically, their property.'

The *Portsmouth Evening News* of Thursday 17 January 1935 reported:

THE SELSEY TRAMWAY CLOSING DOWN THIS WEEK The news that the tramway service between Chichester and Selsey will be definitely closing down at the end of this week will be received with regrets by those who know how useful this service has been since it was started in 1897. Employees of the Tramway Company are under notice to terminate their appointments this week-end. Inquiries reveal that no passengers have been carried this week, and no goods accepted for transport other than those for which invoices had previously been received. For the conveyance of goods it has been of decided value. What is going to happen appears to be obscure at the moment, but the rapid development of the Witterings and the steady growth of Selsey seem to suggest that there is case for railway development along the coast, perhaps from Bognor Regis.

The *Locomotive* for February 1935 reported, 'Chichester & Selsey Light Railway. – On Saturday Jan. 19, the West Sussex Light Railway closed down for all traffic.'

The *Railway Magazine* for March 1935 said that on 19 January the train service was abandoned. The same magazine for April 1935 carried two illustrated articles on the railway, one by R.W. Rush and one by Dr Hugh Nicol.

According to the *Railway Observer* of February 1935, the railway closed on 12 January 1935. Attempts to dispose of the railway to the Southern were unsuccessful and owing to motor bus competition it was forced to close. For some reason it got the date of closure wrong.

## OTHER CLOSURES

The section of the former SECR Greenwich Park branch not used for the Lewisham Loops officially closed in 1932.

Fairfield Sidings – the remains of the former LBSCR Croydon Central branch closed on 14 September 1934

1 January 1934 saw the joint GWR and SR bus service from Weymouth to Wyke Regis transferred to the Southern National Omnibus Company.

In 1934, the SR obtained an Act which authorised it to abandon the Gravesend and Rochester Canal. Only the Gravesend basin remained in use by pleasure boats.

## OPENINGS OF STATIONS

The following stations were opened during the period covered by this chapter:

Syon Lane 5 July 1931
Allhallows-on-Sea 15 May 1932
Bricklayer's Arms goods, formerly also passengers, reopened to
    passengers for summer Sunday excursions in June 1932
Stokes Junction Halt 15 July 1932
Stoneleigh 17 July 1932
Woodmansterne 17 July 1932
Sandsfoot Castle Halt 1 August 1932

Richborough Castle Halt 19 June 1933
Paulsgrove Halt 28 June 1933 – race meetings only
Berrylands 16 October 1933
Albany Park between Sidcup and Bexley 7 July 1935
Bingham Road Halt 30 September 1935 reopened
Coombe Lane 30 September 1935 reopened
Tinsley Green for Gatwick Airport 30 September 1935 – renamed
    Gatwick Airport 1 June 1936

## CLOSURES OF STATIONS

The following stations were closed during the period covered by
this chapter:

Blean and Tyler Hill Halt 1 January 1931
South Street Halt 1 January 1931
Tankerton Halt 1 January 1931
Whitstable (Canterbury and Whitstable) 1 January 1931
Whitstable Harbour (Canterbury and Whitstable) 1 January 1931
Lee on Solent 1 January 1931
Sandgate 1 April 1931
Wherwell 6 July 1931
Bentworth and Lasham – passengers 12 September 1932 – goods
    1 Jun 1936
Cliddesden – passengers 12 September 1932 – goods 1 Jun 1936
Herriard – passengers 12 September 1932 – goods 1 Jun 1936
Shoreham Airport – 1 January 1933 – reopened 1 July 1935 –
    closed 15 July 1940
Bishop's Waltham 2 January 1933 – passengers
Devils Dyke 2 January 1933 – goods – passengers 1 January 1939
Durley Halt – 2 January 1933
Kemp Town 2 January 1933 – passengers
Lewes Road – 2 January 1933 – passengers
Ebbsfleet and Cliffsend Halt 1 April 1933

Rosherville 16 July 1933

Ruthern Bridge 30 December 1933 the last train had run on 29 November

Stone Cross Halt 7 July 1935

Cocking 8 July 1935

Lavant 8 July 1935 – passengers

Axon Lodge private station 30 September 1935

Barnstaple Town Wharf (Pilton Yard) 30 September 1935

Blackmoor 30 September 1935

Bratton Fleming 30 September 1935

Caffyns Halt 30 September 1935

Chelfham 30 September 1935

Hurn 30 September 1935

Lynton and Lynmouth 30 September 1935

Parracombe 30 September 1935

Snapper Halt 30 September 1935

Woody Bay 30 September 1935

## STATION RECONSTRUCTIONS

There were two station reconstructions completed between 1931 and 1935. At Hastings, the station was reconstructed in a neo-Georgian style in 1931 to the design of the architect James Robb Scott. The *Hastings and St Leonards Observer* of Saturday 11 July 1931 reported:

Hastings' handsome new railway station, erected at a cost of over £200,000 on the site the old without an hour's interruption in the normal railway traffic, was opened by the Mayoress (Mrs. G.H. Ormerod) at mid-day Monday (6 July). With a silver key presented by the Chairman of the Southern Railway (Everard Baring) the Mayoress unlocked the large glass-panelled doors of the spacious octagonal booking-hall, and a in brief inaugural speech declared "this beautiful Hastings Station" open.

At Dover, following the consolidation of passenger services, except for those for the continent, the Priory Station was modernised between 1930 and 1932 at a cost of £135,000 and opened on 8 May 1932. The *Dover Express* of 13 May reported:

The new Dover Priory Station was opened on Sunday. There was no ceremony. The new station, with its striking frontage of stone with, in big green- letters, "Southern Railway," and the band of green round the overhanging shelter around the station is altogether different to the old station, and will for years be an outstanding feature of Dover. The station, with its three long platforms, gives far greater accommodation than the old station, and it now possible to originate a train from the station without blocking the main line. Some think the cover only being over the platform gives less protection from the weather than the old station did, but now the accommodation has all been opened to the public there are waiting rooms, heated in the winter, and refreshment rooms, that did not exist before. Inside the station the rooms have a high dado of white tiles with white walls above, concrete floors well finished off, and plenty of light.

## CAR FERRY

In 1931, the SR put into service Britain's first railway-owned cross-Channel car ferry. The ship *Autocarrier* was built and engined by D and W Henderson of Glasgow and was launched on 5 February that year. It was a steel twin-screw steamer with a capacity for 307 passengers and 26 cars and following official trials on 23 March the ship arrived at Dover on 26 March and made its maiden voyage to Calais on 30 March. The *Autocarrier*'s route alternated between Dover and Calais in the summer and Folkestone and Boulogne in the winter and provided competition for Townsend's existing car ferry. Coincidentally Townsend's put a new car ferry on the Dover to Calais route on 30 March 1931. The *Daily Mirror* of 31 March reported:

CAR CHANNEL SERVICE. Motoring Tourist Steamers That Save Time and Worry. Two steamers specially designed to provide motoring tourists with a first-class cross- Channel service made their first voyages of the season yesterday between Dover and Calais, writes our motoring 'correspondent. They provide accommodation which will enable car owners and their friends to travel with the vehicle and thus avoid much of the delay and inconvenience of car transport services on this route in the past. The *Forde*, which belongs to Messrs. Townsend Brothers, has room for about thirty cars and 168 passengers. The *Autocarrier*, a new vessel owned by the Southern Railway, can accommodate thirty-five cars and 120 passengers. The return rates for cars are approximately the same by both vessels, and in each case the single passenger fare is 10s.

## SOUTHAMPTON DOCKS

According to Klapper, during the days of the LSWR the possibilities of extending the quays at Southampton along the river Test towards Millbrook had been detailed by the railway to Southampton Corporation and in 1922 agreement was reached. At the SR's board meeting in October 1923 it was resolved to obtain powers to reclaim the mud flats and the following month to obtain powers to widen the main line between Southampton West and Millbrook. The plan that evolved was, to enclose 407 acres of the mud flats by means of constructing a quay wall and then infilling so as to reclaim the whole area to solid ground. Parallel with the quay wall would be built a jetty 4,500 long and 600 feet away from the quay. As a result of discussions between Walker, Gilbert Szlumper (who was at the time the SR's Docks and Marine Manager), and his father Alfred Szlumper, who was at the time the Southern's Chief Engineer, some bargaining went on. The result was that at the cost of £33,000 to Southampton Corporation, 100 acres of the reclaimed land became the property of the Corporation and the town was given the assurance that Southampton West station would be enlarged from a small two platform roadside station into one with two island platforms and that improvements would be made to the road layout to ease access to the new quays.

The estimated cost of the work was about £13 million and that it would eventually provide an additional 16,600 ft run of quay space. It was July 1926 before the board agreed to adopt Walker's recommendation to carry out the work in three stages. The new works in Stage 1 to be carried on the capital account. The first contracts were let in the autumn of 1926 and was for the dredging of a 35ft channel and provided 3,500ft of quay at which place there would 45ft of water at low tide. It was estimated that the cost of the works at the remodelling of Southampton West station would be £2,896,000 plus a further £30,000 for the junction of the new access railway at Millbrook, which latter had somehow escaped from the listing in the first instance. The whole work was to be supervised by the SR's Docks Engineer, F.E. Wentworth-Shields. The second stage of the work involved extending the quay wall to 7,500ft in length with the additional portion being dredged to give 40ft of water alongside at low water ordinary spring tides. According to Klapper, the entire quay was capable of taking eight of the world's largest liners at any one time. The reclamation of 407 acres of mudflats enabled the SR to pass to Southampton Corporation a large area for municipal purposes and to provide a large industrial estate. The final stage was the construction of a new graving dock for the overhaul of the largest liners. The culmination of the works was a ceremony on 26 July 1933 when King George V declared open and named after himself the world's largest graving dock, which alone had cost £1¾ million.

The *Daily Herald* of 27 July 1933 reported:

KING OPENS GIANT DOCK CHILDREN'S CHEERS FROM DECKS OF LINER Fifty thousand people listened to loud–speakers in Southampton broadcasting the King's speech when he opened the biggest graving dock in the world here to-day. Millions more, all over the world, listened to the broadcast by the BBC. The King and Queen arrived at Southampton from Cowes in the Royal yacht. *Victoria and Albert*, and the first to greet them were Chelsea pensioners, on holiday at Netley Hospital, who were stationed at the end of the Hospital Pier A thousand school children waved white flags from the liner *Empress of Britain* and 4.000 more people filled the grandstand.

The New Docks or Western Docks as they were later known were not actually completed until 1934.

Between 1934 and 1935, Southampton West station was rebuilt in the manner indicated earlier and on 7 July 1935 it was renamed Southampton Central. The new station buildings were largely constructed from concrete and in the then fashionable art deco style

## WATERLOO

According to the *Railway Magazine* of October 1934, although there were eight running lines approaching Waterloo station from Clapham Junction from south to north, they were arranged in the following order: Down main local, Down main through, Up main relief, Up main through, Up main local, Down Windsor local, Down Windsor through, Up Windsor. The magazine said that the fouling of the departure roads by arriving trains that had to cross them and vice versa had handicapped the full use of the 21 platform lines and 13 platforms. The situation had gradually grown worse concurrently with the increase in suburban traffic since the electric services were first introduced in 1916. It had therefore been decided to remedy the evil and to make the approaches to Waterloo accord better with the design and needs of the station. The eight running lines were to be rearranged on the principle of alternating Up and Down lines and part of the big scheme was to provide a flying junction at (Durnsford Road) Wimbledon. In addition there was to be colour light signalling between Waterloo and Hampton Court Junction and a power operated signal box at Waterloo.

The *Locomotive* of February 1935 reported:

Southern Railway.— To improve the facilities for working trains into and out of Waterloo it has been decided to put in hand a big scheme to eliminate the complicated system of cross-over roads between Vauxhall and Waterloo. A "fly-over" will be constructed near Durnsford Road, Wimbledon, to enable a transposition of lines to be made between Wimbledon and Waterloo. The new arrangement will result in the following: – The down local line will remain as at present; the present down through line will become the up local line; the up through line will become the down through line and the up local line will become the up through line. The effect will be that trains from Kingston, Hampton Court, etc., will be taken over from the present up line side to the present down side, giving direct access to their respective platforms at Waterloo, and thus avoiding the cross-overs at present used and releasing the main track for the main line and other trains. The track and cross-overs between Waterloo and Vauxhall will require revision, and also the platforms at Vauxhall station. The scheme also includes the installation of three aspect colour light signalling between Waterloo and Hampton Court Junction. This will involve the replacement of the present manual signal box at Waterloo by an electrically worked box similar to those at London Bridge, Cannon Street, and Charing Cross stations. It is anticipated that when these alterations are completed it will be possible to run trains at intervals of from 2 to 2½ minutes, as compared with the present margin on this section of 4 minutes. At present more than 1,200 trains a day pass in and out of Waterloo Station.

According to the *Railway Magazine* of February 1935 details were now available of the scheme to improve the approach to Waterloo station, the cost of which would be approximately £500,000. The transposition would result in the following: the Down local line would remain as at present, the present Down through line would become the Up local line, the Up through line would become the Down through line, the Up local line would become the Up through line. Trains from Kingston, Hampton Court and so forth would be taken from the present Up lines over to the present Down side giving direct access to their respective platforms at Waterloo. In addition to the work involving the construction of the flyover at Durnsford Road, Wimbledon, the track and the crossovers between Waterloo and Vauxhall would be revised. The present down main local platform at Vauxhall was to be replaced by a new island platform between the Down main local line and what was to become the up main local line. Each running line at Vauxhall would thus have a platform face. The new arrangements would make possible running trains at from 2 to 2½ minute intervals compared to the present 4 minute intervals.

The *Locomotive* of November 1935 reported:

Southern Railway.— In connection with the re-arrangement of the lines between Wimbledon and Waterloo the down local platform at Vauxhall Station was closed from Nov. 3 for about 18 weeks, so that it may be replaced by a new island platform serving both up and down local trains. During the transition period passengers may travel to Waterloo to join their trains or down to Clapham Junction by Windsor line trains and change there. Work is progressing on the fly-over bridge at Wimbledon, a gradient of 1 in 45 being arranged for the approach and 1 in 40 down towards Earlsfield.

Alan A. Jackson in *London's Termini* (David and Charles 1969) says that from the Durnsford Road Wimbledon flyover to Clapham Junction the tracks would be: Down main local, Up main local,

Down main through, Up main through, joined at Clapham Junction by Down Windsor local, Down Windsor through, Up Windsor through, Up Windsor local. From Vauxhall to Waterloo an additional Up main relief road would be sited between the Up main through and the Down Windsor local. As before, there would be only one Up Windsor road on that stretch and at Waterloo platforms 7 to 9 would be electrified and all electric platforms up to 9 would be available to the Up main local. At the same time, colour light signalling, with track circuits and power operated points would be installed between Waterloo and Hampton Court Junction. Jackson adds that engineers assembled and tested the new Waterloo track layout in a field near Mitcham. Bringing the story to a conclusion, the flyover and most of the colour light signalling were brought into use on Sunday 17 May 1936, but the signal changes at Waterloo were not brought into use until 18 October after the end of the summer season. On 5 July the weekday electric services had been rearranged. Under this, Dorking North and Effingham Junction via Leatherhead trains ran fast between Waterloo and Wimbledon off peak and Waterloo and Motspur Park in the peak hours supplemented by an all stations service both in the peak and off peak hours from Waterloo to Motspur Park.

According to the *Daily Herald* of 16 May 1936:

WATERLOO HUSTLE There will be an unusual silence at Waterloo Station between one a m and seven a.m. to-morrow. Trains that normally thunder over the lines will be standing silent in sidings for the first time in history, while 1.000 men work at top speed so that Mr Smith and Mr. Jones may reach the Office two minutes earlier in the morning. Linesmen will be laying new rails and engineers removing four hundred semaphore signals on a job that has taken six months to complete, and is costing the Southern Railway £500.000. A fly-over bridge at Wimbledon Park junction will enable electric trains on the

Waterloo-Surbiton lines to cross each other's paths without either train being stopped.

The same newspaper of 18 May reported:

400 Signals Lose Their Arms. While London slept early yesterday the great change-over on the Southern Railway between Waterloo and Surbiton, was completed. Four hundred signals were stripped of their arms, converting eight miles of track from semaphore to the coloured-light system, and the fly-over track between Wimbledon and Earlsfield, constructed to relieve congestion and speed up travel was brought into use. A new arrangement of tracks outside Waterloo Station was also pronounced well and truly laid by the engineers. By 9 30 a.m. all the down trains were running to schedule, and the up trains were averaging only four to five minutes late—which, according to an official, was "not bad going." One result of the change-over is that the company can now run more trains to cope with increased traffic. Nearly a thousand men were engaged in the task of making the new track and converting the signalling, the cost of which is estimated at £500,000.

The *Railway Magazine* of July 1936 carried an illustrated four page article entitled 'Improving the Approach to Waterloo Station, Southern Railway'.

## THE SOUTHERN TAKES TO THE AIR

Bonavia says that in 1933 the General Managers of the big four railway companies met and decided to form a joint railway airline to be called Railway Air Services (RAS). Klapper says that in April 1933 it came to the knowledge of the SR's board through the lively outlook of John Elliot, who by now was assistant to the traffic manager for the development of traffic, of the GWR's intentions as to the introduction of internal air services. John Stroud in *Railway Air Services* (Ian Allan 1987) says that the GWR officially started a service from Cardiff via Haldon to Plymouth on 11 April 1933, with the service being opened to the public the following day. On 22 May the service was extended to Birmingham (Castle Bromwich). The service lasted until 30 September, when it was suspended for the winter.

According to *The Story of Former UK Carrier Railway Air Services* by Mark Finlay on the website *Simply Flying,* RAS was founded in 1934 when the big four railway companies got together with Imperial Airways. According to him, aircraft had the advantage over trains when it came to travel to Ireland, the Scottish Islands and the Channel Islands, to which one could add the Isle of Man, and that a journey from London to Belfast by train and ship took 11½ hours, whereas by air it only took four. What also could be added would be a journey from Cardiff to Plymouth which would involve travelling via Bristol, whereas an aircraft could fly direct.

Finlay says that in forming the partnership between the railways and Imperial Airways they would not be competing against each other as Imperial Airways supplied everything to do with the flying side of the business including aircraft, pilots, maintenance and airfields, whilst the railways handled the administrative duties, the sales of tickets and marketing.

Of the big four companies only the LNER, although part of RAS never actively participated in it.

Prior to the Second World War, RAS' two main bases were Croydon Airport for London and Liverpool's Speke Airport (now Liverpool John Lennon Airport).

The *News Chronicle* of 22 February 1934 said:

RAILWAYS IN NEW AIR COMPANY £50,000 CAPITAL The four main line railway companies and Imperial Airways, Ltd., have reached agreement for the formation of a new company with a nominal capital of £50,000 to provide and operate air services in the British Isles and elsewhere and to form connecting links with the services of Imperial Airways. The general lines of the new organisation have also been agreed, but no announcement in regard to particular services can be made for some time as a number of preliminary arrangements are first necessary.

Just over a month later, the *News Chronicle* on 27 March reported:

Railway Air Services have not decided what services they will run, but they will concentrate on a main north-south service between London and Scotland, via Manchester or Liverpool, extending to Belfast. The fast planes ordered for them will bring Glasgow within three hours' journey of London.

Exactly two days later the *News Chronicle* said:

3 NEW RAIL-AIR LINE S Summer Plans for Britain The full directorate of Railway Air Services. Ltd.—the combine formed by the railway companies to take full advantage of internal airline development was announced last night. The four group railways and Imperial Airways are represented as follows: Sir Harold Hartley, Chairman (London Midland and Scottish Railway); Lt.-Col. H. Bureball (Imperial Airways, Ltd.); Mr. S.B. Collett (Great Western Railway); Mr. G.H. Corble (London and North Eastern Railway); Mr. G.S. Szlumper (Southern Railway). The Southern Railway have arranged to operate a service with Spartan Air Lines between London and the Isle of Man; the LMS is to run a London-Glasgow-Belfast service, and the G.W.R. propose to operate a Plymouth-Cardiff-Bristol- Birmingham service.

The Isle of Man is what the *News Chronicle* said, even though it clearly means the Isle of Wight.

According to the *Croydon Times* of 18 April:

SOUTHERN AIR SERVICES Isle of Wight In 90 Minutes The first of the Southern Air Services will commence on 1st May – between London and the Isle of Wight. This service has been arranged by the Southern Railway in co-operation with the Spartan Air Lines. Limited, and will be operated entirely by the latter. There will be three services in each direction daily, two in the morning and one in the evening. These services will start from Imperial Airways, Victoria Station – passengers being conveyed by motor-car to Croydon Air Port, and thence by air to the Isle of Wight, landing at Cowes. The over-all time of journey will be 1½ hours, the actual air portion occupying 50 minutes. The service should particularly appeal to London business men. The single fare for the whole journey will be 30s. (return fare 50s). and holders can, if they so desire. return first class by the Southern Railway steamers via Ryde and Portsmouth and restaurant car-expresses to London, without extra charge. The new service will therefore provide, under one charge, the possibility of travelling by air, rail, sea and road.

According to *Bradshaw's International Air Guide for November 1934* the service after leaving Croydon called at Ryde, but only if circumstances permitted, Bembridge and Cowes.

The *Daily Herald* of 30 April had an article which mentioned the air service to Isle of Wight which commenced the next day and also one operated by the LMS from London to Glasgow via Nottingham and Manchester and added that in July the GWR would start a service between Plymouth, Cardiff, Birmingham and Liverpool. The same newspaper of Wednesday 2 May then said that the GWR would start its service between Plymouth, Torquay, Cardiff, Birmingham and Liverpool on the following Monday (7 May).

30 July saw the start of a joint GWR and SR service between Birmingham and Cowes on the Isle of Wight via Bristol and Southampton.

The *News Chronicle* of 20 July said, 'NEW AIR SERVICE An air service bringing Liverpool within three hours of the Isle of Wight will be inaugurated at the end of this month by Railway Air Services, whose twin-engined eight-seater liners will call at Birmingham, Bristol and Southampton en route.' This was something of a misnomer as the *Daily Mirror* of 30 July reported, more correctly:

JOURNEY CUT BY 4 HOURS Midlands—Isle of Wight Air Service With the introduction to-day of the new Birmingham – Bristol – Southampton – Cowes Railway Air Service, there will be a saving of four hours in the journey time between Birmingham and the Isle of Wight.. The quickest time by express train and boat is five hours fifty-five minutes, compared with one hour fifty-five minutes by the new service. An even greater saving of seven hours five minutes will be given [to] Railway Air Service passengers from Liverpool, who will connect with the new service at Birmingham. From Bristol the journey time will be reduced from three hours fifty-three minutes to fifty minutes. Connecting air services will also be given by Western Airways to and from Cardiff at Bristol, to Southampton and the Isle of Wight, and from Birmingham at Bristol to and from Bournemouth. The new service will be operated by De Havilland Dragon eight-seater two-engined aeroplanes. A morning and afternoon service will be given in each direction on week-days. In all cases the aerodromes will be connected with the Southern and Great Western Railway Companies' stations by road motor services which, at Southampton, will be extended to, or run from, any berth in the docks if desired. Air passengers have the privilege of returning by rail if they wish, while holidaymakers holding the return halves of summer or tourist tickets may return by air upon payment of a supplement. Each passenger (children excepted) is allowed hand luggage up to 35 lb., heavy luggage being collected and conveyed by the railway companies under their " luggage in advance system.

The *Weekly Dispatch* of 12 August reported:

> RAIL AEROPLANES The first air excursions to the seaside are to be run next Wednesday by Railway Air Services. They will go from Liverpool to Cowes, Birmingham and Plymouth. and from Birmingham to Plymouth and Cowes Passengers from Liverpool will leave by the 8.45 a.m. 'plane, and the ordinary single fare of 70s, 30s., and 85s. will be charged for the double journey. From Birmingham passengers will leave by the 9.35 a.m. 'plane, and the excursion fares will be 45s. to Cowes and 60s. to Plymouth, compared with the ordinary return fares of 72s. 6d and 90s.

On 20 August a route was started from London to Glasgow via Birmingham, Manchester and Belfast. In the winter Liverpool was substituted for Manchester as RAS was unhappy with the small Barton Aerodrome that served Manchester

A lot of the services did not operate in the winter. Bradshaw's International Air Guide for November 1934 shows only a London to Belfast via Liverpool service operating. All the other services were suspended during the winter i.e. London to the Isle of Wight, Birmingham via Bristol and Southampton to Cowes on the Isle of Wight and Plymouth via Haldon (Teignmouth), Cardiff and Birmingham to Liverpool.

On 27 May 1935, the joint GWR and SR service between Birmingham and Cowes via Bristol and Southampton resumed, but altered to work Liverpool–Birmingham–Bristol–Southampton–Portsmouth–Shoreham. That summer the joint Southern and Spartan Air Lines London to the Isle of Wight service resumed, but with Heston airport rather than Croydon as the London terminus. The service served Bembridge, Cowes and Lea for Sandown and Shanklin. In addition a Lea–Cowes–Southampton was operated under charter to Spartan Air Lines.

Beginning on 28 July, a Southern Sunday air excursion was started between Shoreham and Le Touquet in France. Stroud says that the fares were £3 single and £3 10s day return, which included admission to the Casino at Le Touquet and tea. The service lasted until 1 September.

On 14 September, the Liverpool to Shoreham service ceased for the winter, but the London to the Isle of Wight service continued until it was suspended on 13 January 1936.

## THE BOURNEMOUTH BELLE

The *Bournemouth Belle* was an all-Pullman train which commenced running on Sunday 5 July 1931. The *Richmond Herald* of 27 June 1931 reported:

SUMMER SERVICES ON SOUTHERN RAILWAY. BETTER SUNDAY SERVICES. The Southern Railway's full programme of summer trains will be introduced on Sunday, July 5th, and will remain in operation until Sunday, September 20th. In addition to the usual summer service of trains, many special features will he included. A new first and third class "All Pullman" express will run daily between London, Southampton and Bournemouth, and. will be known as the *"Bournemouth Belle."* This train will leave Waterloo at 10.30 a.m., arriving at Bournemouth Central at 12.39 p.m, Bournemouth West at 12.58 p.m. (Sundays 12.52 p.m.). On week-days only this train will run to Poole, Wareham, Dorchester and Weymouth, due 1.45 p.m. The return service will leave Weymouth at 4 p.m. and Bournemouth Central at 5.10 p.m., due Waterloo at 7.18 p.m.

The same newspaper of 5 September reported:

SOUTHERN RAILWAY. The winter; train service will be introduced on Sunday, September 20th. Extra trains run during the summer to cope with the heavy holiday traffic will

**The all-Pullman** Bournemouth Belle first entered service in June 1931 and with the Southern, later Brighton, Belle was one of two all Pullman trains on the Southern Railway. The train is seen here behind Lord Nelson class No 862 *Lord Collingwood* in its original condition. *R.C. Stumpf Collection*

be discontinued. The principal main line expresses will be retained –, but the *"Bournemouth Belle"* Pullman Limited will run on Sunday until further notice at 10.50 a.m. from Waterloo instead of daily as in the summer months.

On 1 January 1936 the train which ran only all the week in the summer and only on Sundays in the winter, commenced to run daily throughout the year. The News Chronicle of 2 January 1936 reported

Luxury Train's New Ways The luxury express, *Bournemouth Belle*, yesterday embarked on a new stage of its bright career.

Since this Pullman train was introduced in July, 1931, she has run only during the summer months and on Sundays in the winter. She now becomes a daily feature of Southern Railway travel between Waterloo, Southampton and Bournemouth.

## PRESERVATION AND CENTENARIES

Starting in 1932 there was an attempt that was initially successful to preserve the oldest surviving former Isle of Wight Railway 2-4-0 side tank No W13 *Ryde* built in 1864. The locomotive was withdrawn on 2 July 1932 and great efforts were made on the Island to gain permanent preservation, but money was scarce. In June 1934, *Ryde* was transferred to the paint shop at Eastleigh Works, where according to D.L. Bradley in *A Locomotive History of Railways on the Isle of Wight* (RCTS 1982), several other aged engines had been collected to form the nucleus of a SR museum. The story does not have a happy ending as in 1940 during the Second World War the locomotive was broken up with hardly a voice raised in protest.

The *Locomotive Magazine* of March 1933 reported:

The Railway and Travel Correspondence Society has opened a subscription list with the object of preserving the veteran Southern Ry. tank locomotive W13 *Ryde*. This engine was one of three built by Beyer, Peacock & Co. in 1864 for the Isle of Wight Ry., when it was opened, and at the time of its withdrawal in 1932 was the oldest passenger locomotive in service in the country. It is proposed to recondition the engine in the pregrouping colours and preserve it on the Island. The hon. secretary of the society, Mr. Geo. R. Grigs, 31 Belsize Road, N.W.6., will receive subscriptions.

The Railway and Travel Correspondence Society opened its appeal in the *Railway Observer* of February 1933, but unfortunately, according to the *Railway Observer* of October 1933, the appeal was somewhat disappointing and that one last appeal was being made after which nothing further is heard of this in the *Railway Observer*.

1934 saw the centenary of the oldest section pf the former LSWR, which was opened from Bodmin to Wadebridge on 4 July 1834 and from Boscarne to Wenford Bridge and from Grogley Halt to Ruthern Bridge on 30 September 1834. The *Railway Magazine* for July and August and October had a three-part illustrated article on the history of the railway by Charles E. Lee and in addition the magazine for October also carried details of the centenary celebrations held to commemorate the event. On 5 September, a party of railway and press representatives travelled from London and Exeter to Wadebridge by the 10 35 am portion of the *Atlantic Coast Express*, the Padstow portion of which called there at 4 40 pm. Following a reception on the station platform, the various exhibits that had been arranged by the joint initiative of the SR, the local authorities and the Old Cornwall Society were inspected. In the station yard a SR third class coach and a restaurant car of the latest type were drawn up. After this, a visit was paid to the Town Hall where various Bodmin and Wadebridge relics were assembled, which included the terminal milestone found at Ruthern Bridge, a section of rail still mounted on its original granite sleeper, a guard's horn lantern and a wooden pattern used for casting the wheels of the rolling stock, as well as documents, tickets and photographs relating to the railway. In the evening a journey was made to Bodmin and back hauled by Class O2 0-4-4 side tank locomotive No 216 with its smokebox decorated with flags. The exhibition at Wadebridge was open to the public on 5 and 6 September with nearly a thousand visitors passing through the modern rolling stock on the first day alone. At Waterloo Station in London, where the first and second class Bodmin and Wadebridge carriage of 1834, was permanently exhibited in the station concourse, it appeared blushingly amid a halo of flags. The SR issued special cheap tickets to Wadebridge from within a radius of sixty miles. A special six penny return was available over the Bodmin and

Wadebridge branch, which the *Railway Magazine* said represented a fare of less than a halfpenny a mile for the journey in both directions. The *Western Moring News* of 6 September reported:

THREE TRAINS A WEEK Cornwall's Pioneer Railway TRIALS OF TRAVEL 100 YEARS AGO Centenary Celebrated At Wadebridge TRIBUTES to Cornish railway pioneers and to the men who showed considerable foresight years ago were paid yesterday, when the centenary of the Bodmin-Wadebridge railway was observed. First there was a "Cornish Tea" in Wadebridge Town Hall, which was attended by Southern Railway officials, various representative people, members of local government bodies, and a party of Press representatives. A trip was made over the old line from Wadebridge to Bodmin and back, and this morning there will be a motor tour of the district, after which the railway officials and London journalists will leave Wadebridge at 12.50 p.m. for Waterloo. The present line follows the same route as the original, but at one or two places the old curves have been straightened. On either side is beautiful country, green with the recent refreshing rains, and heather-covered slopes, impressive in their ruggedness. Sitting in carriages fitted with all modern conveniences and comforts, the passengers gazed upon this scene while traversing one of the most famous railway lines the world, a line which will go down in history as one the earliest railways in the country, and certainly the first in the West of England.

On one occasion (in the early days of the Railway) a commercial traveller came running to the station a few minutes after the train had passed on. When is the next train?" he demanded. The cool reply Mr. Samuel Worth, whose name will always be remembered in the annals of railway history as first station-master of the line, was "At five o'clock the day after tomorrow." Passenger trains ran only on Tuesdays. Thursdays and Saturdays in those days.

Mr. Ernest Martin, chairman Wadebridge Old Cornwall Society, presided at the tea, and said his father, who as a schoolboy attended the cutting of the first sod of the railway, began his business career in the office of the company at Wadebridge. An exhibition in connection with the centenary was staged in Wadebridge Town Hall, and included many interesting photographs, relics, prints and old tickets, which were a source of delight to members of Old Cornish Societies and people interested in local history.

The *Locomotive Magazine* of October 1934 carried an article on the centenary of the Bodmin and Wadebridge Railway:

Bodmin-Wadebridge Railway Centenary. CENTENARY celebrations took place on Wednesday, Sept. 5, at Bodmin and Wadebridge to commemorate the opening of the Bodmin & Wadebridge Ry. In Wadebridge Town Hall an exhibition of relics was staged. Photographs and engravings, old maps and documents relating to its history were on view, together with a lantern which used to be carried by the guard, old tickets, a sleeper to which a piece of old rail adheres, and a milestone recording the distance from Wadebridge to Ruthern Bridge in miles, furlong, chains and yards. Representatives of public bodies, the Southern Rly and various associations were invited to a Cornish tea at Wadebridge after which Mr. Ernest Martin, chairman of the Old Cornwall Society, who presided, spoke with praise of the pioneers, whose work was being remembered. First a canal to link the rivers Fowey and Camel was proposed, but when that was found to be impracticable because of the expense and intervening hills, a railway was projected. Sir William Molesworth, squire of Pencarrow, was the leader of the project and became chairman of the company. "This railway taught Stephenson how to make railways," said Mr. Martin, who told how it was first intended to haul trains by horses from Dunmere up the steep gradient to Bodmin. "But on the eve of the opening the engine driver was asked to try to get the engine up the incline. He tried and succeeded in getting in all three wagons up to Bodmin. A railway representative then dashed off post haste to Glasgow to tell George Stephenson about the achievement and that was the means of shortening the line from Edinburgh to Glasgow then about to be constructed." In a siding at Wadebridge the Southern Rly exhibited a modern dining car and a third class coach. During the evening the railway and press representatives travelled up the line to Bodmin where an old milestone in the wall near the railhead and a fragment of the old line which used to enter a private store— long before the railway track at the station was lowered—were inspected. There were also present four surviving members of the staff of the old line, before it was connected up to the North Cornwall section of the L&SWR in 1888, the oldest being 82. For over 60 years the Bodmin–Wadebridge line, even though belonging for a long period of that time to the L.&S.W.R., remained isolated from the greater system by many miles.

## ACCIDENTS

The years 1931 to 1935 saw two major accidents on the SR. The first was at Raynes Park on 25 May 1933 when a passenger train was derailed, coming to rest foul of an adjacent line. Another passenger train was in a sidelong collision with it. Five people were killed and thirty-five were injured. The cause of the accident was a failure to implement a speed restriction on a section of track that was under maintenance.

The *Daily Herald* of 26 May reported:

FIVE KILLED IN LONDON TRAIN SMASH. Express Grinds Deadly Way Along Line of Derailed Carriages. DRIVERS' FIGHT

TO AVERT COLLISION THEY FORESAW. Five passengers were killed, ten taken to hospital injured and a number of others less seriously hurt in a sideways crash between an express and a slow steam train on the Southern Railway between Wimbledon and Raynes Park SW, yesterday. With both drivers striving gallantly to avert the collision they foresaw, the slow train rocked off the metals and the express running by its side in the opposite direction, splintered its way down the line of coaches. Rescuers, mobilised hurriedly from all around, worked heroically. Some of the injured tended those who were more hurt than themselves and all panic was avoided.

The second accident occurred on 4 September 1934, when two freight trains collided at. There were no fatalities.

The *News Chronicle* of 5 September reported:

TRAIN SMASH: FIVE HURT Waggons Piled Up to Bridge Parapet An engine-driver was seriously injured and four other railwaymen hurt in a collision between two goods trains at Hither Green station yesterday. One of the trains was backing out of a siding. It consisted of several fruit vans and some of these were smashed and were forced on to the top of other waggons. On the other train a waggon laden with stone was lifted on to the waggon in front and one end rested on the parapet of a bridge. Traffic was stopped on the up and down lines to Orpington. George Bishop (60). of Walthrop Avenue. Tottenham. is in Lewisham Hospital with an injury to the head. A guard named Frank Vousden had an arm broken. and three other men had slight injuries.

## MOTIVE POWER

Commencing in 1931 steam locomotives were renumbered as follows. Those whose numbers were prefixed with an E signifying Eastleigh had the prefix dropped. Those whose numbers were prefixed with an

A signifying Ashford had their numbers increased by 1,000, whilst those whose numbers were prefixed by a B signifying Brighton had their numbers increased by 2,000. Those whose numbers began with 0 had it changed to 3 and those on the Isles of Wight kept their W prefix.

There was only one class of steam locomotive build during the period of this chapter and that was the W Class three-cylinder 2-6-4 side tank. According to Bradley, the SR's electrification of its suburban lines whilst removing the problems that had plagued the Central and Eastern sections for several decades created new ones, principally the integration of the slower and less flexible long distance passenger services and local goods services. The difficulties of the former were overcome by dexterous timetable compilation. The problem of the latter remained with short distance and cross London goods services by necessity having to run them during the day when the density of the electric services was at their greatest. The solution was a three-cylinder goods version of the Class K1 three-cylinder 2-6-4 side tank. The necessary drawings were prepared at Ashford Drawing Office and were presented to the Locomotive Committee for comment and acceptance on 25 March 1929. Because of the recent rebuildings of Class K and K1 some scepticism was expressed, but Maunsell assured the Committee that the locomotives would not be used on passenger trains and the Class known as Class W was approved. The first five locomotives were built at Eastleigh Works in January and February 1932, whilst a further ten were built at Ashford Works from April 1935 to April 1936. The locomotives incorporated a few parts from the K and K1 Class locomotives including tanks, steps, bogies and parts of the coal bunkers. The cabs, which boasted a side window, were new, whilst the bogie wheels had steam brakes to compensate for the lack of a tender. The locomotives proved a success and were well liked by the engine crews.

Class N No A816 was experimentally fitted with a steam heat conservation system devised by the Scottish marine engineer A.P.H. Anderson. The locomotive entered Eastleigh on 12 July 1930, with conversion starting on 16 August. The locomotive finally left the works for trials on 8 August 1931. With an array of pipes, a tank on each of the running plates and a square chimney the locomotive was not an attractive sight. Trials plus a few modifications including the provision of a normal round chimney were conducted up until October 1933 when the locomotive was put into store until 27 May 1935 when the equipment was removed and the locomotive re-entered traffic on 3 August 1935.

Between 16 October 1933 and 3 February 1934 Class N No 1850 was experimentally fitted with a valve gear devised by T.J. Marshall of Harrogate. The trials against Class N No 1413 fitted with normal Walschaerts did not offer any advantage and the experimental valve gear was removed.

With the Eastbourne and Hastings electrification there was no longer any work for the former LBSCR L Class 4-6-4 side tank locomotives and the decision was made to rebuild them as 4-6-0 tender locomotives. The rebuilding was carried out between December 1934 and April 1936. Previous to that Maunsell had considered having the locomotives rebuilt as 4-6-0s following the Sevenoaks accident. The new class number for the locomotives was N15X. The Locomotive for December 1934 recorded:

The seven well-known express passenger tank engines of the 4-6-4, or "Baltic" type, originally designed for the former London, Brighton & South Coast Railway by Mr. L. B. Billinton and built at that Company's Brighton Works between the years 1914 and 1922, are being converted by Mr. R..E.L. Maunsell, chief mechanical engineer of the Southern Railway, into 4-6-0 type tender engines at Eastleigh Works, and the first to be completed is No. 2329, Stephenson.

It has to be said that whilst the locomotives in their original form had been very good as rebuilt, they were rather mediocre and were

**Following the** completion of the electrification to Eastbourne and Hastings there was no useful work for the L Class 4-6-4Ts and they were all rebuilt as N15X Class 4-6-0s. In their new guise the locomotives were not as successful as they were in their original form. *R.C. Stumpf Collection*

nowhere as good as either Urie or Maunsell's N15s and after an initial spell on top link duties were relegated to spending their time on secondary duties.

In November 1933, a design was produced at Eastleigh for a four-cylinder 4-6-2. The Civil Engineer accepted the design and granted it the same route availability as the Lord Nelsons but with a speed restriction of 80mph. Maunsell said that the class was discarded because of the high cost at £13,750 per engine and tender and the small number required. In 1934-5, a design was proposed for a four-cylinder 2-6-2 locomotive; the Civil Engineer restricted its use to the lines from Waterloo to Bournemouth and Victoria to Dover via Tonbridge and a maximum speed of 70mph and in consequence nothing came of this. It is not quite sure if this was to be a purely passenger locomotive or a mixed traffic locomotive. In late 1935 there was an abortive proposal for a four-cylinder 4-6-0. According to *Locomotive Magazine* of March 1927 there was a proposal to build some 4-8-0 goods engines with the same boiler, motion, etc as the Lord Nelsons. They were intended for heavy goods work – principally from the Kent coalfield and it was intended to build seven of the locomotives with the numbers 1881 to 1887 being allocated to them. The reason the design was rejected was that trials with doubled headed coal trains had proved that the running loops on the main line, sidings and some block sections were too short for such trains. According to the Civil Engineer, £61,000 would have had to be spent at once and a further £52,000 later before full use could be made of the class. D.L. Bradley in Part 1 of *Locomotives of the Southern Railway* (RCTS 1975) gives the date as 1935. Even more interesting was a proposal for a 4-6-2 + 2-6-4 Beyer Garratt, which was originally proposed for the Continental expresses on the Eastern Section but which the Civil Engineer would only pass to work between Basingstoke and Exeter. Each engine unit was to have three cylinders. Whilst a change of locomotive at Exeter on the West Country trains was the norm, having a change at Basingstoke defeated the object.

During the early 1930s, smoke deflectors were fitted to locomotives of the King Arthur (N15), Lord Nelson, Schools (V), N, N1, U and U1 classes as well as the N15X Class.

## ELECTRIC MULTIPLE UNITS

For main line electrification up to the end of 1935, the following classes of emus were built.

4LAV – four-car non-gangwayed units which had one trailer composite carriage in which there was a lavatory at each end. These units were built in 1931-2 for stopping services to Brighton

6PUL – six-car units gangwayed within the unit and included a composite Pullman car. They were built for express service to Brighton.

6CITY – very similar but were built for the London Bridge to Brighton City Limited business expresses but had more first class accommodation than the 6PUL units. Both 6PUL and 6CITY were built 1931-2.

5BEL – five-car Pullman car units comprising of first and third class Pullman cars and were built in 1932 for the *Southern,* later *Brighton Belle*.

2NOL – two-car non-gangwayed units using former LSWR carriage bodies on new frames and were used on slow services on the South Coast and in South London. As their name implies, they did not possess lavatory accommodation. They were built in 1934-5.

2BIL – two-car units were first built in 1935 for the Hastings and Eastbourne electrification and although non-gangwayed had a lavatory within each carriage.

6PAN – similar to the 6PUL but the Pullman car was replaced by a Pantry car. They were built in 1935 for express service to Hastings and Eastbourne.

## RAILBUS

The railbus was a Sentinel-Cammell steam railbus ordered in October 1932 and delivered in March 1933. It was clearly a success, but when Maunsell contacted Sentinel for a quote for a further five, the quote that came back for each railbus was somewhat in excess of the amount paid for the original one. According to Chacksfield, the original quote had been aimed in the hope that a follow-up order was a foregone conclusion. So no; Maunsell was no fool!

## SHIPS

On the Isle of Wight services, the paddle steamer PS *Sandown* was built in 1934 by William Denny and Brothers of Dumbarton for the Portsmouth to Ryde service. MV *Hilsea* was built in 1931 by William Denny and Brothers of Dumbarton for the Portsmouth Harbour to Fishbourne car ferry service:

SS *Isle of Sark* was built in 1932 and the SS *Brittany* was built in 1933, both by William Denny and Brothers of Dumbarton for services from Southampton to Le Havre / the Channel Islands / St Malo.

SS *Brighton* was ordered from William Denny and Brothers of Dumbarton in 1932 and delivered in 1933 for the Newhaven to Dieppe service which was one third owned by the Southern and two thirds owned by the State Railway of France:

SS *Autocarrier* was Britain's first purpose built car ferry. The ship was built in 1931 by D, and W. Henderson of Glasgow for the Dover to Calais and Folkestone to Boulogne services:

# LOCOMOTIVES, SHIPS AND BUSES 1923 TO 1949

## LOCOMOTIVES

| Class | Company of origin | Designer | Wheel arrangement | First built | Date of last withdrawal | Notes |
|---|---|---|---|---|---|---|
| A | LCDR | Kirtley | 0 4 4 t | 1875 | 1926 | |
| A1 | LCDR | Kirtley | 0 4 4 t | 1880 | 1926 | |
| A1 | LBSCR | Stroudley | 0 6 0 t | 1872 | 1963 | |
| A1X Rebuilt of A1 | LBSCR | Marsh | 0 6 0 t | 1911 | 1963 | |
| A2 | LCDR | Kirtley | 0 4 4 t | 1880 | 1926 | |
| A12 | LSWR | Adams | 0 4 2 | 1887 | 1948 | |
| B | SER | Stirling | 4 4 0 | 1898 | 1931 | |
| B1 | SECR Rebuilt of B | Wainwright | 4 4 0 | 1910 | 1951 | |
| B1 | LBSCR | Stroudley | 0 4 2 | 1882 | 1933 | |
| B1 | LCDR | Kirtley | 0 6 0 | 1877 | 1924 | |
| B2 | LCDR | Kirtley | 0 6 0 | 1891 | 1933 | |
| B2X Rebuilt of B2 | LBSCR | Marsh | 4 4 0 | 1907 | 1933 | B2 first built 1895 |
| B4 | LSWR | Adams | 0 4 0 t | 1891 | 1963 | |
| B4 | LBSCR | Billinton R | 4 4 0 | 1899 | 1951 | |
| B4X | LBSCR Rebuilt of B4 | Billinton L | 4 4 0 | 1922 | 1951 | |

| Class | Company of origin | Designer | Wheel arrangement | First built | Date of last withdrawal | Notes |
|---|---|---|---|---|---|---|
| B12 | GER | Holden S D | 4 6 0 | 1911 | 1961 | LNER worked US Army ambulance trains on SR in 1944. |
| C | SECR | Wainwright | 0 6 0 | 1900 | 1967 | |
| C1 | LBSCR | Stroudley | 0 6 0 | 1882 | 1925 | |
| C2 | LBSCR | Billinton R | 0 6 0 | 1893 | 1950 | |
| C2X rebuilt of C2 | LBSCR | Marsh | 0 6 0 | 1908 | 1962 | |
| C3 | LBSCR | Marsh | 0 6 0 | 1906 | 1952 | |
| C8 | LSWR | Drummond | 4 4 0 | 1898 | 1938 | |
| C14 | LSWR Rebuilt under Urie | Drummond/ Urie | 0 4 0 t/ 2 2 0 t | 1906 built as 2 2 0t rebuilt as 0 4 0 t 1913-23 | 1959 | |
| D | SECR | Wainwright | 4 4 0 | 1901 | 1956 | |
| D1 | SECR Rebuilt of D | Maunsell | 4 4 0 | 1921 | 1962 | |
| D1 | LBSCR | Stroudley | 0 4 2t | 1873 | 1951 | |
| D1X rebuilt of D1 | LBSCR | Marsh | 0 4 2t | 1910 | 1933 | |
| D3 | LBSCR | Billinton R | 0 4 4 t | 1892 | 1955 | |
| D3X Rebuilt of D3 | LBSCR | Marsh | 0 4 4 t | 1909 | 1948 | |
| D15 | LSWR | Drummond | 4 4 0 | 1912 | 1956 | |
| E | SECR | Wainwright | 4 4 0 | 1905 | 1955 | |
| E1 | SECR Rebuilt of E | Maunsell | 4 4 0 | 1919 | 1962 | |
| E1 | LBSCR | Stroudley | 0 6 0t | 1875 | 1961 | |

| Class | Company of origin | Designer | Wheel arrangement | First built | Date of last withdrawal | Notes |
|---|---|---|---|---|---|---|
| E1X Rebuilt of E1 | LBSCR | Marsh | 0 6 0t | 1911 | 1948 | |
| E1R | SR rebuilt of LBSCR | Maunsell | 0 6 2t | 1927 | 1959 | |
| E2 | LBSCR | Billinton L | 0 6 0t | 1913 | 1963 | |
| E3 | LBSCR | Stroudley / Billinton R | 0 6 2t | 1891 | 1959 | |
| E4 | LBSCR | Billinton R | 0 6 2t | 1897 | 1963 | |
| E4X Rebuilt of E4 | LBSCR | Marsh | 0 6 2t | 1909 | 1959 | |
| E5 | LBSCR | Billinton R | 0 6 2t | 1902 | 1956 | |
| E5X Rebuilt of E5 | LBSCR | Marsh | 0 6 2t | 1911 | 1956 | |
| E6 | LBSCR | Billinton R | 0 6 2t | 1904 | 1962 | |
| E6X Rebuilt of E6 | LBSCR | Marsh | 0 6 2t | 1911 | 1959 | |
| E10 | LSWR | Drummond | 4 2 2 0 | 1901 | 1927 | |
| E14 | LSWR | Drummond | 4 6 0 | 1907 | 1959 as rebuilt to H15 1914 | |
| F | SER | Stirling | 4 4 0 | 1883 | 1930 | |
| F1 | SECR Rebuilt of F | Wainwright | 4 4 0 | 1903 | 1949 | |
| F6 | LSWR | Adams | 0 4 4 t | 1894 | 1951 | |
| F9 | LSWR | Drummond | 4 2 4t | 1899 | 1940 | |
| F13 | LSWR | Drummond | 4 6 0 | 1905 | 1961 as H15 rebuilt 1924-5 | |
| G | SECR | Pickersgill | 4 4 0 | 1899 To traffic 1900 | 1927 | Originally built for GNOSR |
| G6 | LSWR | Adams | 0 6 0 t | 1894 | 1961 | |

| Class | Company of origin | Designer | Wheel arrangement | First built | Date of last withdrawal | Notes |
|---|---|---|---|---|---|---|
| G14 | LSWR | Drummond | 4 6 0 | 1908 | 1925 | |
| G16 | LSWR | Urie | 4 8 0t | 1921 | 1962 | |
| H | SECR | Wainwright | 0 4 4 t | 1904 | 1964 | |
| H1 | LBSCR | Marsh | 4 4 2 | 1905 | 1951 | |
| H2 | LBSCR | Marsh | 4 4 2 | 1911 | 1958 | 32424 last 4 4 2 in service in Britain |
| H15 | LSWR | Urir | 4 6 0 | 1914 | 1961 | |
| H15 | SR rebuilt of LSWR | Maunsell | 4 6 0 | 1924 | 1961 | |
| H15 SR new build | SR | Maunsell | 4 6 0 | 1924 | 1961 | |
| H16 | LSWR | Urie | 4 6 2 t | 1921 | 1962 | |
| I1 | LBSCR | Marsh | 4 4 2 t | 1906 | 1932 last one rebuilt to I1X | |
| I1X Rebuild of I1 | SR | Maunsell | 4 4 2 t | 1925 | 1951 | |
| I2 | LBSCR | Marsh | 4 4 2 t | 1907 | 1939 | |
| I3 | LBSCR | Marsh | 4 4 2 t | 1907 | 1952 | |
| I4 | LBSCR | Marsh | 4 4 2 t | 1908 | 1940 | |
| J | SECR | Wainwright | 0 6 4 t | 1913 | 1951 | 31596 last 0 6 4t in Britain |
| J1 | LBSCR | Marsh | 4 6 2 t | 1910 | 1951 | |
| J2 | LBSCR | Marsh | 4 6 2 t | 1912 | 1951 | |
| K | LBSCR | Billinton L | 2 6 0 | 1913 | 1962 | |
| K | SECR | Maunsell | 2 6 4 t | 1 in 1917 plus more in 1925-6 | 1928 rebuilt to U | |
| K1 | SR | Maunsell | 2 6 4 t | 1925 | 1928 rebuilt to U1 | |
| K10 | LSWR | Drummond | 4 4 0 | 1901 | 1951 | |
| K14 | LSWR | Drummond | 0 4 0t | 1908 | 1959 | |

| Class | Company of origin | Designer | Wheel arrangement | First built | Date of last withdrawal | Notes |
|---|---|---|---|---|---|---|
| KES | KESR | Hawthorn Leslie | 0 8 0 t | Built 1904 Acquired 1932 | 1950 | |
| L | SECR | Wainwright/ Maunsell | 4 4 0 | 1914 | 1961 | |
| L | LBSCR | Billinton L | 4 6 4 t | 1914 | 1936 Last one as 4-6-4t | |
| L1 | SR | Maunsell | 4 4 0 | 1926 | 1962 | |
| L11 | LSWR | Drummond | 4 4 0 | 1903 | 1952 | |
| L12 | LSWR | Drummond | 4 4 0 | 1904 | 1955 | |
| Leader | SR | Bulleid | 0 6 6 0 t | 1949 | 1951 | |
| LN | SR | Maunsell | 4 6 0 | 1926 | 1962 | |
| M1 | LCDR | Kirtley | 4 4 0 | 1880 | 1923 | |
| M2 | LCDR | Kirtley | 4 4 0 | 1884 | 1923 | |
| M3 | LCDR | Kirtley | 4 4 0 | 1891 | 1928 | |
| M7 | LSWR | Drummond | 0 4 4 t | 1897 | 1964 | |
| MN | SR | Bulleid | 4 6 2 | 1941 | 1967 | |
| N | SECR | Maunsell | 2 6 0 | 1917 | 1966 | |
| N1 | SECR | Maunsell | 2 6 0 | 1923 | 1962 | |
| N15 | LSWR | Urie | 4 6 0 | 1918 | 1958 | |
| N15 SR new build | SR | Maunsell | 4 6 0 | 1925 | 1962 | |
| N15X SR rebuild of L | SR rebuilt of LBSCR | Maunsell | 4 6 0 | 1934 | 1957 | |
| O | SER | Stirling | 0 6 0 | 1878 | 1932 | |
| O1 | SECR Rebuilt of O | Wainwright | 0 6 0 | 1903 | 1961 | 31065 last former SER locomotive in service |
| O2 | LSWR | Adams | 0 4 4 t | 1889 | 1967 | |
| O4 | LSWR | Adams | 0 4 2 | 1893 | 1948 | 30629 last 0 4 2 in service in Britain |

| Class | Company of origin | Designer | Wheel arrangement | First built | Date of last withdrawal | Notes |
|---|---|---|---|---|---|---|
| P | SECR | Wainwright | 0 6 0 t | 1909 | 1962 | |
| P14 | LSWR | Drummond | 4 6 0 | 1910 | 1927 | |
| Q | SER | Stirling | 0 4 4 t | 1881 | 1929 | |
| Q | SR | Maunsell | 0 6 0 | 1938 | 1965 | |
| Q1 | SECR Rebuilt of Q | Wainwrigt | 0 4 4 t | 1903 | 1930 | |
| Q1 | SR | Bulleid | 0 6 0 | 1942 | 1966 | |
| R | LCDR | Kirtley | 0 4 4 t | 1891 | 1955 | 31666 last LCDR loco in service |
| R | SER | Stirlling | 0 6 0t | 1888 | 1943 | |
| R1 | SECR Rebuilt of R | Wainwright | 0 6 0t | 1910 | 1960 | |
| R1 | SECR | Kirtley | 0 4 4 t | 1900 | 1956 | |
| S | SECR Rebuilt of C | Maunsell | 0 6 0 st | 1917 | 1951 | |
| S11 | LSWR | Drummond | 4 4 0 | 1903 | 1955 | |
| S15 | LSWR | Urie | 4 6 0 | 1920 | 1964 | |
| S15 SR new build | SR | Maunsell | 4 6 0 | 1927 | 1966 | |
| S160 | USATC | Marsh Major J W US Army | 2 8 0 | 1942 | 1944 | Transferred to Continent after D Day 1944 |
| T | LCDR | Kirtley | 0 6 0 t | 1879 | 1952 | |
| T1 | LSWR | Adams | 0 4 4 t | 1888 | 1936 | |
| T3 | LSWR | Adams | 4 4 0 | 1892 | 1945 | |
| T6 | LSWR | Adams | 4 4 0 | 1895 | 1943 | |
| T7 | LSWR | Drummond | 4 2 2 0 | 1897 | 1927 | |
| T9 | LSWR | Drummond | 4 4 0 | 1899 | 1961 | |
| T14 | LSWR | Drummond | 4 6 0 | 1911 | 1951 | |

| Class | Company of origin | Designer | Wheel arrangement | First built | Date of last withdrawal | Notes |
|---|---|---|---|---|---|---|
| U Rebuild of K and new | SR | Maunsell | 2 6 0 | 1928 | 1966 | |
| U1 Rebuild of K1 and new | SR | Maunsell | 2 6 0 | 1928 | 1963 | |
| USA Ex USATC S100 | SR | Hill Colonel H G US Army | 0 6 0 t | 1942 Date to service 1946 | 1967 | |
| V | SR | Maunsell | 4 4 0 | 1930 | 1962 | |
| W | SR | Maunsell | 2 6 4 t | 1932 | 1964 | |
| WC/BB | SR | Bulleid | 4 6 2 | 1945 | 1967 | |
| WD 2-8-0 | Wd | Riddles | 2 8 0 | 1944 | 1950 On Southern | |
| X2 | LSWR | Adams | 4 4 0 | 1890 | 1942 | |
| X6 | LSWR | Adams | 4 4 0 | 1895 | 1946 | |
| X14 | LSWR | Drummond | 0 4 4 t | 1911 | 1964 | |
| Z | SR | Maunsell | 0 8 0t | 1929 | 1962 | |
| 700 | LSWR | Drummond | 0 6 0 | 1897 | 1962 | |
| 2301 | GWR | Dean | 0 6 0 | 1883 | 1957 | Used on EVLR in the Second World War |
| 046 | LSWR | Adams | 4 4 2t | 1879 as 4 4 0t rebuilt 1883 | 1925 | |
| 0135 | LSWR | Adams | 4 4 0 | 1880 | 1925 | |
| 0273 | LSWR | Beattie | 0 6 0 | 1872 | 1924 | |
| 0302 | LSWR | Beattie rebuilt Adams | 0 6 0 | 1874 Rebuilt 1886 | 1935 | |

| Class | Company of origin | Designer | Wheel arrangement | First built | Date of last withdrawal | Notes |
|---|---|---|---|---|---|---|
| 0329 | LSWR | Beattie rebuilt Adams and Urie | 2 4 0 wt | 1874 Rebuilt 1886 and 1921 | 1962 | Last survivors of a class first built in 1863 |
| 0330 | LSWR | Beattie | 0 6 0 st | 1876 | 1933 | |
| 0380 | LSWR | Adams | 4 4 0 | 1879 | 1925 | |
| 0395 | LSWR | Adams | 0 6 0 | 1881 | 1959 | |
| 0415 | LSWR | Adams | 4 4 2 t | 1882 | 1961 | |
| 0445 | LSWR | Adams | 4 4 0 | 1883 | 1926 | |
| 0460 | LSWR | Adams | 4 4 0 | 1884 | 1929 | |
| IOW | IOWR | Beyer Peacock | 2 4 0 t | 1864 | 1933 | |
| IOWC | IOWCR | Beyer Peacock | 2 4 0 t | 1876 | 1929 | |
| IOWC | IOWCR | Black Hawthorn | 4 4 0 t | 1880 | 1926 | |
| IOWC A1X | IOWCR ex LBSCR | Stroudley rebuilt Marsh | 0 6 0 t | 1872 as A1. Rebuilt to A1X 1916-1930 | 1963 | |
| FYN | FYNR | Manning Wardle | 0 6 0 st | 1902 | 1932 | |
| FYN | FYNR ex LBSCR A1 | Stroudley | 0 6 0 t Rebuilt as A1X 1932 | 1877 | 1963 | |
| L&B | L&B | Manning Wardle | 2 6 2 t | 1897 | 1935 | |
| L&B | L&B | Baldwin | 2 4 2 t | 1900 | 1935 | |
| SR | SR | Manning Wardle | 2 6 2t | 1925 | 1936 | Withdrawn 1935, but used on demolition train into 1936. |
| PDSWJ | PDSWJ | Hawthorn Leslie | 0 6 0t | 1906 | 1951 | |
| PDSWJ | PDSWJ | Hawthorn Leslie | 0 6 2t | 1907 | 1957 | |
| SR | SR | Macleod | 0 4 0 hand powered | 1930 | 1938 | |

| Class | Company of origin | Designer | Wheel arrangement | First built | Date of last withdrawal | Notes |
|---|---|---|---|---|---|---|
| SECR RM | SECR | Wainwright | 0 4 0 t +4 | 1905 | 1924 | |
| SR RB S | SR | Sentinel | 0 4 0 t +4 | 1933 | 1940 | |
| SR RB P | SR | Drewry | 4 w | 1927 | 1934 | |
| D3/12 | SR | Maunsel | 0 6 0 de | 1937 | 1964 | |
| D3/13 | SR | Bulleid | 0 6 0 de | 1949 | 1971 | |
| 11001 | SR | Bulleid | 0 6 0 dm | 1949 | 1959 | |
| D16/2 | SR | Bulleid | 1 Co Co 1 | 1950 | 1963 | |
| CC1 | SR | Bulleid | Co Co | 1941 | 1968 | |
| W&C | LSWR | Drummond | Bo | 1898 | 1968 | |
| Shunter | LSWR | Drummond | Bo Bo | 1899 | 1965 | |
| W&C 4 car | LSWR | Drummond | 4 car | 1898 | 1940 | |
| W&C1 car | LSWR | Drummond | 1 car | 1899 | 1940 | |
| SR W &C units | SR | Bulleid | 5 car | 1940 | 1993 | |
| 2 SL | LBSCR | Marsh rebuilt Maunsell | 2 car | 1908/ 1929 | 1954 | |
| 2 WIM | LBSCR | Marsh rebuilt Maunsell | 2 car | 1911/ 1929 | 1954 | |
| 2NOL | SR | Maunsell | 2 car | 1934 | 1959 | |
| 2 HAL | SR | Bulleid | 2 car | 1938 | 1971 | |
| 2 BIL | SR | Maunsell | 2 car | 1935 | 1971 | |
| 3 SUB | LSWR/SR | Urie/ Maunsell | 3 car | 1916 | 1949 Last conversion to 4SUB | Some converted from ac 4/5 car stock built 1925 |
| 4 SUB | SR | Bulleid | 4 car | 1941 | 1983 | Includes conversions from 3SUB |
| 4DD | SR | Bulleid | 4 car | 1949 | 1971 | |
| 4 LAV | SR | Maunsell | 4 car | 1930 | 1969 | |
| 4 COR | SR | Maunsell/ Bulleid | 4 car | 1937 | 1971 | |

| Class | Company of origin | Designer | Wheel arrangement | First built | Date of last withdrawal | Notes |
|---|---|---|---|---|---|---|
| 4 BUF | SR | Maunsell | 4 car | 1938 | 1971 | |
| 4 RES | SR | Maunsell | 4 car | 1937 | 1971 | |
| 5 BEL | SR | Maunsell | 5 car | 1932 | 1972 | |
| 6 CIT | SR | Maunsell | 6 car | 1932 | 1969 | |
| 6 PAN | SR | Maunsell | 6 car | 1932 | 1969 | |
| 6 PUL | SR | Maunsell | 6 car | 1932 | 1969 | |
| Ryde | Ryde | Ryde electric | 1 car | 1886 | 1927 | |
| Ryde | SR | MaunselL petrol/ diesel | 1 car | 1927 | 1969 | |
| 27A | SDJR | George England | 2 4 0 | 1861 | 1925 | |
| 28A | SDJR | George England | 2 4 0 st | 1861 | 1928 | |
| 1 | SDJR | Fox Walker | 0 6 0 st | 1874 | 1934 | |
| 8 | SDJR | Fox Walker | 0 6 0 | 1876 | 1928 | |
| 9 | SDJR | Johnson | 0 4 4 t | 1877 | 1946 | |
| 14 | SDJR | Johnson | 4 4 0 | 1891 | 1932 | |
| 19 later 18 | SDJR | Fowler, Leeds | 0 6 0 | 1874 | 1928 | |
| 19 | SDJR | Fowler | 0 6 0 t | 1929 | 1967 | |
| 23 | SDJR | Johnson | 0 6 0 | 1881 | 1933 | |
| 57 | SDJR | Fowler | 0 6 0 | 1922 | 1965 | |
| 62 | SDJR | Johnson | 0 6 0 | 1896 | 1962 | |
| 67 | SDJR | Johnson rebuilt Fowler | 4 4 0 | 1895 | 1962 | |
| 77 | SDJR | Deeley | 4 4 0 | 1908 | 1938 | |
| 80 | SDJR | Fowler | 2 8 0 | 1914 | 1962 | |
| 86 | SDJR | Fowler | 2 8 0 | 1925 | 1964 | |
| 101 | SDJR | Sentinel | 0 4 0 t | 1929 | 1961 | |
| 25A | SDJR | SDJR | 0 4 2 st | 1885 | 1928 | |
| 45A | SDJR | SDJR | 0 4 0 st | 1895 | 1929 | |

# SHIPS

| Name | Company of origin | Builder | Date into service | Date out of service |
|---|---|---|---|---|
| SS *Alberta* | LSWR | Clydebank Engineering and Shipbuilding Company | 1900 | 1930 sold |
| SS *Ardena* | LSWR | A McMillan and Son, Dumbarton 1915<br><br>Rebuilt Caledon Shipping and Engineering Company | 1920 | 1934 |
| SS *Arromanches* | SR/SNCF | Forges et Chantiers de La Medeiterranen Le Havre | 1946 | 1964 |
| SS *Arundel* | LBSCR | William Denny and Brothers, Dumbarton | 1900 | 1934 |
| SS *Autocarrier* | SR | D and W Henderson, Glasgow | 1931 | 1952 |
| SS *Biarritz* | SECR | W Denny & Bros, Dumbarton 1914 war service | 1921 | 1948 |
| SS *Brighton* | LBSCR | W Denny & Bros, Dumbarton | 1903 | 1944 |
| SS *Brighton* | SR | W Denny, Dumbarton | 1933 | 1940 sunk |
| SS *Brittany* | LSWR | Earle's Co Ltd, hull for LBSCR sold 1912 to LSWR | 1910 for lbscr 1912 for lswr | 1936 |
| SS *Brittany* | SR | William Denny and Brothers, Dumbarton | 1933 | 1962 |
| SS *Caesarea* | LSWR | Cammell Laird | 1910 | 1923 |
| SS *Canterbury* | SECR | William Denny and Brothers, Dumbarton | 1900 | 1926 |
| TSS *Canterbury* | SR | William Denny and Brothers, Dumbarton | 1929 | 1964 |
| SS *Cherbourg* | LSWR | Not known | 1873 | 1939 |
| SS *Deal* | SR | D & W Henderson & Co, Glasgow | 1928 | 1963 |
| SS *Dieppe* | LBSCR | Fairfield Shipbuilding & Engineering Co Ltd, Govan, Glasgow | 1905 | 1944 |
| SS *Dinard* | SR | W Denny, Dumbarton | 1924 | 1958 |
| PS *Duchess of Albany* | LBSCR/LSWR JOINT | Scotts, Greenock | 1889 | 1928 |

| Name | Company of origin | Builder | Date into service | Date out of service |
|---|---|---|---|---|
| PS *Duchess of Kent* | LBSCR/LSWR JOINT | Day, Summers and Company | 1897 | 1933 |
| PS *Duchess of Fife* | LBSCR/LSWR JOINT | Clydebank Engineering and Shipbuilding Company | 1899 | 1929 |
| PS *Duchess of Norfolk* | LBSCR/LSWR JOINT | D & W Henderson & Co, Glasgow | 1911 | 1937 |
| SS *Empress* | SECR | William Denny and Brothers, Dumbarton | 1907 | 1923 |
| SS *Engadine* | SECR | William Denny and Brothers, Dumbarton | 1911 | 1932 |
| TSS *Falaise* | SR | William Denny and Brothers, Dumbarton | 1947 | 1974 |
| SS *Fratton* | SR | D & W Henderson & Co, Glasgow | 1925 | 1944 |
| PS *Freshwater* | SR | John Samuel White & Co Cowes | 1927 | 1959 |
| SS *Hampton Ferry* | SR | Swan Hunter on the Tyne | 1934 | 1969 |
| SS *Hantonia* | LSWR | Fairfield, Govan | 1911 | 1952 |
| SS *Haslemere* | SR | D & W Henderson & Co, Glasgow | 1925 | 1959 |
| SS *Hythe* | SR | D & W Henderson & Co, Glasgow | 1925 | 1956 |
| SS *Invicta* | SECR | William Denny and Brothers, Dumbarton | 1905 | 1923 |
| SS *Invicta* | SR | D & W Henderson & Co, Glasgow<br><br>Built 1940 but requisitioned by government | 1945 | 1972 |
| SS *Isle of Guernsey* | SR | William Denny and Brothers, Dumbarton | 1930 | 1969 |
| SS *Isle of Jersey* | SR | William Denny and Brothers, Dumbarton | 1930 | 1959 |
| SS *Isle of Sark* | SR | William Denny and Brothers, Dumbarton | 1932 | 1960 |
| SS *Isle of Thanet* | SR | William Denny and Brothers, Dumbarton | 1925 | 1963 |

| Name | Company of origin | Builder | Date into service | Date out of service |
|---|---|---|---|---|
| SS *Laura* | LSWR | Aitken and Mansel, Whiteinch | 1885 | 1928 |
| SS *Londres* | SR/SNCF | Forges et Chantiers de La Medeiterranen Le Havre | 1947 | 1963 |
| SS *Lorina,* | LSWR | William Denny and Brothers, Dumbarton 1918<br><br>Rebuilt Caledon Shipping and Engineering Company | 1920 | 1940 sunk |
| SS *Maid of Kent* | SR | W Denny and Brothers, Dumbarton | 1925 | 1940 |
| SS *Maid of Orleans* | SECR | W Denny & Bros, Dumbarton 1918 War Service | 1920 | 1939 |
| SS *Maid of Orleans* | SR | William Denny and Brothers, Dumbarton 1918<br><br>Rebuilt Caledon Shipping and Engineering Company | 1949 | 1975 |
| SS *Maidstone* | SR | D & W Henderson & Co, Glasgow | 1926 | 1953 southern 1958 irish sea |
| PS *Merstone* | SR | Caledon & Co at Dundee | 1928 | 1950 |
| SS *Minster* | SR | D & W Henderson & Co, Glasgow | 1924 | 1940 |
| SS *Newhaven* | LBSCR/ C FETAT | Forges et Chantiers de La Medeiterranean Le Havre | 1911 | 1940 sunk |
| SS *Normannia* | LSWR | Fairfield, Govan | 1911 | 1940 sunk |
| SS *Paris* | LBSCR | William Denny and Brothers, Dumbarton | 1913 | 1940 |
| PS *Portsdown* | SR | Caledon & Co at Dundee | 1928 | 1941 sunk |
| SS *Princess Ena* | LSWR | Gourlay Brothers, Dundee | 1905 | 1935 sunk |
| PS *Princess Margaret* | LBSCR/LSWR JOINT | Scotts, Greenock | 1893 | 1928 |
| SS *Ringwood* | SR | D & W Henderson & Co, Glasgow | 1926 | 1959 |
| SS *Riviera* | SECR | William Denny and Brothers, Dumbarton | 1911 | 1932 |
| SS *Rouen* | LBSCR | Forges et Chantiers de La Medeiterranean Le Havre | 1912 | 1940 |
| PS *Ryde* | SR | William Denny and Brothers, Dumbarton | 1937 | 1967 |

| Name | Company of origin | Builder | Date into service | Date out of service |
|---|---|---|---|---|
| PS *Sandown* | SR | William Denny and Brothers, Dumbarton | 1934 | 1965 |
| PS *Shanklin* | SR | J.I. Thorneycroft at Southampton | 1924 | 1951 |
| SS *Shepperton Ferry* | SR | Swan Hunter on the Tyne | 1934 | 1974 |
| PS *Southsea* | SR | Fairfield Shipbuilding & Engineering Co Ltd, Govan, Glasgow | 1930 | 1941 Sunk |
| SS *St Briac* | SR | William Denny and Brothers, Dumbarton | 1924 | 1939 |
| SS *Tonbridge* | SR | D & W Henderson & Co, Glasgow | 1924 | 1940 |
| SS *Twickenham Ferry* | SR | Swan Hunter on the Tyne | 1934 | 1972 |
| SS *Vera* | LSWR | Clydebank | 1898 | 1933 |
| SS *Vera* | LSWR | Clydebank | 1898 | 1933 |
| SS *Versailles* | LBSCR | Forges et Chantiers de La Medeiterranean Le Havre | 1921 | 1940 |
| SS *Victoria* | SECR | William Denny and Brothers, Dumbarton | 1906 | 1928 |
| SS *Whitstable* | SR | D & W Henderson & Co, Glasgow | 1925 | 1959 |
| PS *Whippingham* | SR | Fairfield Shipbuilding & Engineering Co Ltd, Govan, Glasgow | 1930 | 1962 |
| SS *Worthing* | SR | William Denny and Brothers, Dumbarton | 1928 | 1960 |
| SS *Alpha* | SDJR | William Swan & Son | 1877 | 1925 |
| SS *Julia* | SDJR | Mordey, Carney & Co, Southampton | 1904 | 1934 |
| SS *Radstock* | SDJR | J Crichton & Co, Saltney and Wrexham. Ship launched from Saltney yard, Chester | 1925 | 1934 |

# BUSES

| Registration no | Builder | Original owner | Date into service | Date out of service |
|---|---|---|---|---|
| CX1683 | Karrier | LSWR | 1914 | 1924 |

**The former** LCDR T Class 0-6-0Ts comprised ten members and were built at the Company's Longhedge Works between 1879 and 1893. No A600, later 1600 was originally LCDR No 141 and was sold out of service in 1940 and survived until 1958 as the last former LCDR locomotive. *John Scott-Morgan collection*

**Class R1 0-4-4T** were an improved version of the LCDR R Class and all were ordered by and delivered to the South Eastern and Chatham Railway. No 1697 is seen here on a passenger working composed of typical SECR stock. *H.C. Casserley*

**At the** formation of the Southern Railway there were 31 members of the former SER O Class 0-6-0s designed by James Stirling still in their original condition with the safety valve in the middle of the boiler and the trademark Stirling style of cab. *Photomatic*

**The SECR** J Class was the Southern Railway's only class of 0-6-4Ts and were all built in 1913 to the design of Harry S Wainwright. The Class all survived into British Railways days and were the last 0-6-4Ts to operate on the British mainland. *Photomatic*

*Above*: **The London,** Brighton and South Coast Railway was the principal constituent company of the Southern Railway to use the 0-6-2T. There were four classes – E3, E4, E5 and E6. E3 No 165 formerly named *Blatchingdon* is seen on a local goods working. The locomotive survived until 1959, *John Scott-Morgan collection*

*Below*: **The London,** Brighton and South Coast Railway unlike the Southern Railway's other constituent companies had only one class of 0-4-4Ts – Class D3. An unidentified member of the class is seen on a local passenger working. *H.C. Casserley*

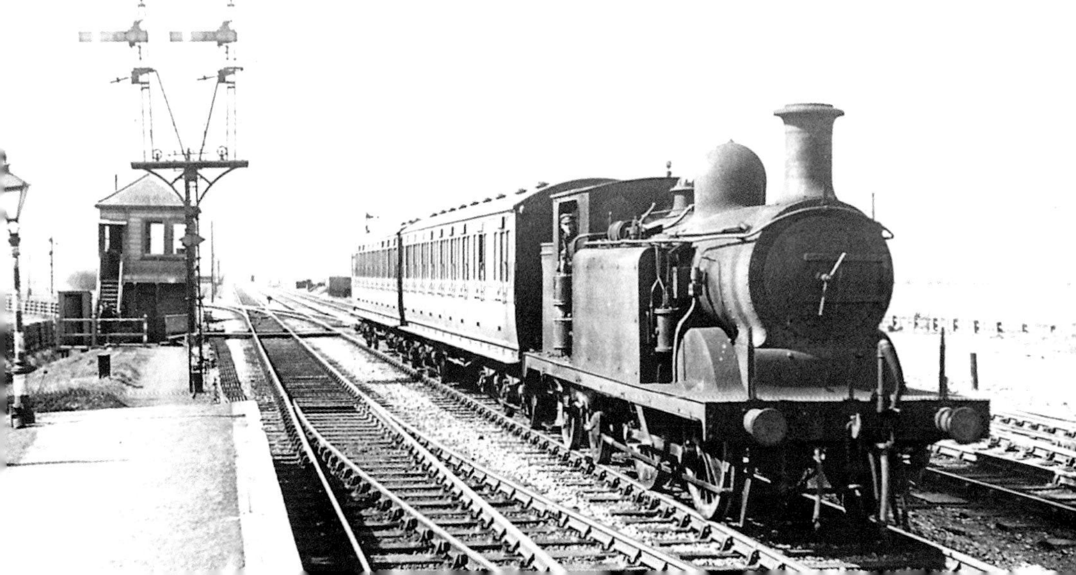

**The London** and South Western Railway class of 10 Class C14 0-4-0Ts were built as 2-2-0Ts to work push-pull trains, but were not a success and four rebuilt to 0-4-0Ts. Seven were sold out of service prior to the grouping. No 3741 is seen during the 1930s. *R.C. Stumpf Collection*

# SOURCES


Allen Cecil J., *Titled Trains of Great Britain* Ian Allan 1967]

Barker T.C. and Robbins, Michael A., *History of London Transport* Allen and Unwin 1963 and 1974

Bradley, D.L., *The locomotives of the South Eastern Railway.* RCTS 1963

Bradley, D.L., *The locomotive history of the London Chatham and Dover Railway.* London: RCTS, 1979.

Bradley, D.L., *The locomotive history of the South Eastern and Chatham Railway.* rev. ed. 1980

Bradley, D.L., *A locomotive history of the railways on the Isle of Wight.* RCTS, 1982.

Bradley, D.L. *The locomotives of the London and South Western Railway.* Solihull: RCTS, 1965. 2v.

Bradley, D.L., *The Locomotives of the London Brighton & South Coast Railway* RCTS 1969-74 3 v

Bradley, D.L., *Locomotives of the Southern Railway.* London, RCTS 1975. 2v.

Bonavia, Michael, *The History of the Southern Railway*, Unwin Hyman 1987

Chacksfield, J.E. *Richard Maunsell : an engineering biography*, OPC 1998

Cummings, Michael, *Railway Motor Buses and Bus Services in the British Isles, 1902-33* OPC 1983

Dendy, Marshall C.F.A., *History of the Southern Railway*, Ian Allan 1963

Elliot, Sir John, *On and Off the Rails*, George Allen and Unwin 1982

Glover, John, *Southern Electric* Ian Allan 2001

Greenway, Ambrose, *A Century of Cross Channel Passenger Ferries,* Ian Allan 1981

Horne, M.A.C., *London's District Railway*, Capital 2018-9 2v

Jackson, Alan A., *London's Termini*, David and Charles 1968

Jackson, Alan A., *London's Local Railways*, David and Charles 1978

Jackson, Alan A., *London's Metropolitan Railway*, David and Charles 1986

Jackson, Alan A., *Semi-Detached London*, Wild Swan 1991

Jackson, Alan A. and Croome, Desmond F., *Rails Through the Clay*, Capital 1993

Kitchenside, Geoffrey, *Railway Carriage Album*, Ian Allan 1986

Kidner, Roger, *The Southern Railway*, Oakwood 1958

Klapper, Charles, *Sir Herbert Walker's Southern Railway*, Ian Allan 1973

Moody, G.T., *Southern Electric* Ian Allan 1968

Snell, John, *One Man's Railway*, David and Charles 1993

Stroud, John, *Railway Air Services*, Ian Allan 1987

Thomas, David St John, *A Regional History of the Railways of Great Britain – The West Country*, David and Charles 1966

White, H.P., *A Regional History of the Railways of Great Britain: Southern England*, Phoenix 1961

White, H.P., *A Regional History of the Railways of Great Britain: Greater London*, Phoenix 1963

White, H.P., *Forgotten Railways: South East England*, David and Charles 1976


*Bradshaw's Railway Guide*

British Newspaper Archive

The Time Digital Archive

British Newspaper Archive

*The Locomotive*

The *Railway Magazine*
The *Railway Observer*

The Great Eastern Railway Society whom I wish to thank for making the *Locomotive* available in digital form.
The Railway Correspondence and Travel Society for making available the Railway Observer online.

# INDEX

Dear Reader,

We hope you have enjoyed this book, but why not share your views on social media? You can also follow our pages to see more about our other products: facebook.com/penandswordbooks or follow us on X @penswordbooks

You can also view our products at www.pen-and-sword.co.uk (UK and ROW) or www.penandswordbooks.com (North America).

To keep up to date with our latest releases and online catalogues, please sign up to our newsletter at: www.pen-and-sword.co.uk/newsletter

If you would like a printed catalogue with our latest books, then please email: enquiries@pen-and-sword.co.uk or telephone: 01226 734555 (UK and ROW) or email: uspen-and-sword@casematepublishers.com or telephone: (610) 853-9131 (North America).

We respect your privacy and we will only use personal information to send you information about our products.

Thank you!